Integrated Assessment Models of Climate Change Economics

Zheng Wang · Jing Wu · Changxin Liu
Gaoxiang Gu

Integrated Assessment Models of Climate Change Economics

Springer

Zheng Wang
Institute of Policy and Management
Chinese Academy of Sciences
Beijing
China

Jing Wu
Institute of Policy and Management
Chinese Academy of Sciences
Beijing
China

Changxin Liu
Beijing
China

Gaoxiang Gu
Population Research Institute
East China Normal University
Shanghai
China

ISBN 978-981-10-3943-0 ISBN 978-981-10-3945-4 (eBook)
DOI 10.1007/978-981-10-3945-4

Library of Congress Control Number: 2017932428

Printed on acid-free paper

This Springer imprint is published by Springer Nature
The registered company is Springer Nature Singapore Pte Ltd.
The registered company address is: 152 Beach Road, #21-01/04 Gateway East, Singapore 189721, Singapore

Preface

Global change is a challenge that mankind faces. Therefore, tackling global change is an important task of scientists. I am a geographer and I have been working on China's historical climate change issue for a long time. We have a unique advantage in this study because of the vast history of China. However, in 1999, I gradually realized the importance of tackling climate change, and China as a superpower should play a greater role in the study. I began to study the problem of global climate change economics according to the requirements of Chinese Academy of Sciences in 2007; further, I found that this is a complex scientific problem combined with physical science and economic science. At this time, the published paper of Prof. Nordhaus and Prof. Yang at AER in 1996 lits up me like a lighthouse, through which I feel that the core problem is IAM.

The global economic crisis took place in 2008 when China was facing two problems: on the one hand, actively involved in tackling global climate change, which the Chinese government put forward the "energy saving and emission reduction" policy; on the other hand, any country's "energy saving and emission reduction" measures are likely to affect other countries and the world economy under the background of economic integration. The reduction measures of multi-countries economic interactions need to be studied facing the global economic crisis. But at this time, all the IAMs I have studied have no economic interaction among countries, and therefore we need to do new exploration. In 2010 we introduced Mundell-Fleming mechanism and technology advances into the popular RICE model to construct MRICE (multifactor RICE), and its first application is the calculation of emission reduction effect of Sino-US economic interaction in a global common emission reduction, which was published in Economic Modeling. Since the simulation requires software development, my graduates Lili Cui, Yihong Jiang, Yiping Zheng, Huaqun Li, Huanbo Zhang, Gangqiang Li, and Jing Wu have been taking part in the work. Jing Wu eventually wrote MRICES software system using C#. At then I pay a visit to Prof. Nordhaus, who gave a friendly reception to me and my assistant, answered some of my questions, and presented me the book of him and Dr. Boyer. In 2012, after improving the characterization of technological progress,

Jing Wu, Shuai Zhang, and I completed MRICES-2012, which were released as a public software.

In 2012, I was fortunate to know Prof. Zili Yang. Common scientific understanding and the affection as Chinese linked us together. We had meaningful discussions and he suggested us to focus on mixed emission reduction and game theory. After 2012, we received a joint support from basic scientific research of Ministry of Science and Technology of China and Chinese Academy of Science, and completed the study on EMRICES in 2014. During this study, my graduates Qianting Zhu, Changjiang Shao, Rui Huang, and Changxin Liu took part in this work. As Jing Wu is the backbone of the first phase of the study, Changxin Liu is the backbone of the second phase. Compared to MRICES-2012, carbon trading analysis, sea level rise, and carbon tax impact analysis are included in EMRICES. Unfortunately, due to various reasons, the impact analyses of sea level change, carbon tax, and pollution tax are developed only in China's module in EMRICES, although it is theoretically possible in each economy.

Both MRICES and EMRICES include the keyword RICE to label that it is developed on the basis of RICE. There is a Chinese proverb, "when you drink water, never forget the man who digs the well." MRICES and EMRICES use of the word RICE to express our respect and gratitude to Prof. Nordhaus and Prof. Yang.

CIECIA in this book is another system we developed which is funded by the basis science research project of Ministry of Science and Technology of China. For the development of this system, we visited Prof. Caldeira at Stanford University, and he discussed the algorithm of the carbon cycle model. CIECIA model for depicting the technological progress and industrial structure evolution introduced the mechanism of evolutionary economics. The global economic system is based on global model from Dr. K.Y. Jin's paper published at AER in 2013 combining with our country economic interaction model. In principle, it is a global general equilibrium model, reflecting the global economic integration, so it is more suitable for studying global carbon governance issues. We hope this model can lead to more scholars' interests to global climate change governance under innovation and global economic integration.

The authors thank the consistent support of academician Yihui Ding of Chinese Academy of Engineering, academician Guanhua Xu, and academician Qun Lin of Chinese Academy of Sciences, and Prof. Shiyuan Xu from East China Normal University, commissioner Tongsan Wang of Chinese Academy of Social Sciences for the work, and we also want to thank Prof. Nordhaus, Prof. Yang, and Prof. Caldeira for their help. Thanks Springer for publishing the book.

The work is supported by major research project of Ministry of Science and Technology of China and carbon special research projects of Chinese Academy of Sciences.

Beijing, China
January 2016

Zheng Wang

Contents

Chapter 1
Introduction

1.1 Integrated Assessment Model of Climate Change and Economy

Integrated Assessment Model of Climate Change, short for IAM, consider various factors fully and comprehensively such as climate, economy and energy use at a global level, its application value and potential have been widely recognized. A large number of literature attempts to explain "Integrated Assessment (IA)", including Weyant et al. (1996). Rotmans et al. (1990)'s definition of IA is widely quoted, IA is a process of combining, interpreting, connecting knowledge from different scientific disciplines; In this process, all causal connection of the problem can be comprehensive evaluated from two aspects: compared with single discipline evaluation, research results of IA have value increment and provide useful information for decision makers. Therefore, IA is about global issues such as climate change, information sharing process repeatedly contacting knowledge (science) and action (policy). Weyant et al. (1996) defines the three goals of the IA: assessment of climate change control policy; unify multiple dimensions of climate change to the same framework; quantify the relative importance of climate change in other human facing environmental and non-environmental fields.

The definitions of IAM in academic are not unified, Weyant et al. (1996) defined IAM as model of any using multidisciplinary research knowledge; Schneider (1997) argued that IAM usually contains a series of sub models from other areas, and is used in the integrated assessment of environmental science, technology and policy issues; Kelly and Kolstad (1999) defined that IAM combing the natural sciences and economics in climate change issue, to evaluate policy options under climate change; Tol (2002) argued that IAM is a multidisciplinary cross model including any physics, chemistry, ecology, economics and politics together;

Z. Wang et al., *Integrated Assessment Models of Climate Change Economics*, DOI 10.1007/978-981-10-3945-4_1

Ackerman et al. (2009) argued that IAM a multi-disciplinary calculation model to study climate change, to assess profit and loss of climate policy using the GCM results. Above all, IAM is multidisciplinary cross large-scale model combined climate model with economic model, for the purpose of studying the climate issue and evaluate climate policy.

Due to IAM has good decision support function, there is a mistaken understanding considering IAM as policy instrument. This is because that IA is not to provide help for government decision-making, but built for solving the problems of the real world, usually, these problems are from multiple multi-disciplinary cross to inter-disciplinary integration development (Schneider 1997). IA is an analysis model about the objective reality with the characteristic of physics and analysis model with economic purpose. IA is not to make decision, but the objective and logical estimation.

1.2 The Classification of IAM

IAM can be divided into different categories. The classification of different scholars is on different starting points. This classification can help us to better understand and compare the differences of IAMs, and identify their function. van Vuuren et al. (2011) thought some IAMs more focused on the economy, such as the integration of multi-sectoral computable general equilibrium model with climate module, these models focus on cost-benefit analysis; Other IAMs mainly focused on the integration of physical process of natural systems and economics (integrated model structure or biophysical effects model), this classification distinguishes IAMs according to the carbon cycle and the description of the temperature change.

In fact the simplicity degree of the carbon cycle and climate systems depends on modeling purpose. For these IAMs who focus on cost-benefit analysis, taking DICE, FUND and MERGE for example, the carbon cycle and climate system have been simplified a lot compared to GCM. The amount of atmospheric carbon dioxide is a function of carbon emissions, and the other greenhouse gas emissions are a fixed invariants. Concentration is directly used to calculate the radiation pressure. Equilibrium temperature changes with the change of the radiation pressure. While the IAMS focused on physical process model pay attention to the climate and the expression of carbon cycle in more detail. Many IAMs use the energy balance model of the bottom-up model with a global carbon cycle model to describe the global climate change and greenhouse gases, such as MAGICC. In addition, there are also using the grid size parameters to drive the agricultural growth model of grid level. There are other IAMs that introduced the terrestrial carbon sink, carbon source on the grid scale, to obtain the more complex relationship of the climate, the carbon cycle, land cover and land use change, such as IMAGE (Bouwman et al. 2006).

Goodess et al. (2003) divided IAM into three categories, IAM based on cost-benefit analysis; IAM based on biophysical and IAM based on policy guidance.

(1) Cost-benefit analysis IAM for policy optimization, such as CETA, DICE, FUND, ICAM-3, MERGE, and the MiniCAM. These models firstly care about the economic consequences of climate change, such as comparing costs for climate change adaption and emissions reduction to assess possible alternative policies. In these models, climate modules are under 2 dimensions, some even are 0 dimension. The calculating of these models is short time-consuming, no more than a few hours. As a result, they can be used to rapidly evaluate emissions reduction agreement, such as Kyoto Protocol.
(2) Biophysical-impact based IAM for policy evaluation, such as CLIMPACTS, ESCAPE, IMAGE and IGSM. These models are more focused on quantitative evaluation of the biophysical rather than economic policy evaluation. They tend to be analyze at the regional level, some analysis can also be integrated into the global level. The advantage of these models is to analyze the impact of climate change on the high spatial resolution. But the disadvantage of these models is that the economic module is relatively weak. These model cannot build the economic relations on the corresponding spatial resolution. Economic module often contains only GDP, population and energy use.
(3) Policy guidance IAM, such as ICLIPS. It transfers economic losses (plants, agriculture, water resources) module through climate impact response function into tolerable windows. Tolerable window is generally expressed by the rise of temperature, rainfall and sea level rise level (Fussel et al. 2003). These restrictions are input into greenhouse gas emissions-climate change module to calculate carbon emissions that can keep consistent with tolerate window Bruckner et al. (2003). This model can be used to calculate the threshold value of climate change.

According to model methodology, Yang (2008) divided IA model of climate change into three categories: a computable general equilibrium model, inter-temporal optimization model, and simulation model.

(1) computable general equilibrium model, such as EPPA model of MIT and SGM model from the Pacific northwest laboratory. CGE models usually use social accounting matrix (SAM) as database to establish the model. It can divide departments and regions in detail, and study the regional economic relations of inter-departments and inter-regions. CGE can provide very useful information when studying future GHG and evaluation strategy of GHG reductions, Modelers can set up a specific structure or module in CGE for the analysis of economic problems. A disadvantage of CGE is that its dynamic characteristic is limited due to the limitation of data. Usually CGE is static or dynamic recursion. At present, there is no CGE model of "visionary" (Yang 2008).
(2) the inter-temporal optimal model, such as RICE from Yale and MERGE. Dynamic or inter-temporal optimization model currently are not elaborate to the

department level. But compared to the CGE model, it has better flexibility in depicting individual decision-making and response to the future events. it is more reasonable than the mechanism of CGE on the inter-temporal optimization. In addition, its dynamic structure is more transparent than CGE.

(3) the simulation model, such as ICAM model of Carnegie Mellon university and IMAGE model from the Netherlands national institute of public health. Scenario simulation model does not need to spend time to find the optimal solution. The entire model is without any decision-making or individual economic optimization behavior. Modeling structure also often take the bottom-up model. Also, the model often lack of connection between economic departments. Economic modules are not usually present in the framework of general equilibrium.

Van Vuuren (2006) divided IAM into three categories: multi-sector general equilibrium model, aggregate general equilibrium model; integrated structure model. His classification is similar to Yang's.

(1) Multi-sector general equilibrium, such as AMIGA, EU-PACE, EPPA, SGM, WIAGEM.
(2) Aggregate general equilibrium, such as MERGE, GRAPE.
(3) Integrated structural model, such as IMAGE, MESSAGE, AIM, MiniCAM.

According to the coupling tightness between economic module and climate module, Bahn et al. (2006) divided IAM into two kinds, one kind is economy, climate and damage module highly merging model, such as RICE, DICE, and MERGE. This kind of model usually takes a long time to find the optimal emissions reduction policy. The other type is IGSM model. Economic system adopts multiple regional general equilibrium model, the climate system adopted high resolution general climate system. But system between the economy and climate subsystem is too simple. The economic system only does damage assessment based on temperature rise, but the development of the economic system itself is not affected (Bahn et al. 2006).

If we continue to thinking in depth following Bahn's opinion, it can be found that currently these classifications ignored a very important point-the influence of climate change on the economic development path.

In economic growth theory, economic growth path draws lots of attention, especially for what factors can affect economic growth. Lucas (2002) pointed out that capital, labor, and technology is the root cause of economic growth, and labor force growth, technological progress is the source of the economic growth. However, climate change has caused negative effects on the economy, actually have affected economic development path. The global climate change bring about economic losses at the first phase, such losses will also affect the investment at the next phase, thus affect the amount of capital in the production function. Nordhaus (2008) took CO2 as a new factor in the production function, like capital, is a kind of inputs, but this factor's influence on economic growth is negative. Therefore, whether brings loss of climate change into economic growth process, determines whether a model recognizes climate change's effect economy development. In fact, almost all of the models considered the economic consequences of the temperature rise. But not all models will make the response of economic loss on the of the economy's trajectory. If according to whether IAM model considering the impact of climate change on economic growth path. Models can be roughly divided into two categories. Models considering the effects of climate change on economic growth path are RICE, DICE, MERGE, WITCH on behalf of the optimization. The mechanism of the model is simple, the physical module is inferior to IMAGE, the economic module is inferior to CGE model, and WIAGEM. But they build the impact of climate change into the economic development module, perhaps this is the charm of DICE/RICE can constantly get developed and attention.

Table 1.1 shows the statistics of some international IAM. Temperature feedback on economic system is an important indicator, it reflects the model's ability to carry the impact of climate change internalized to the economic development path. On the issue of climate change, the biggest advantage of CGE is to set emissions reduction measures at the department level, and can reflect the influence on the various departments under emission reduction measures. GREEN and G-CUBED model, as an early energy environmental CGE model, introduced the calculation of carbon emissions. But there is no climate module. It can be used to calculate the influence of carbon tax and carbon trade on emission reductions. CGE models such as FUND and Wiagem brought damage function of temperature rise, but did not introduce temperature factors in the production function (Kemfert 2002). FUND model, although the loss can be divided into 15 kinds of types, but only shows the temperature rising losses, instead of a given temperature on the influence of the production. The defects of RICE model are obvious, although many other models extended RICE including the economy module and the climate modules, often lost its most essential thing, which is the impact of climate change on economic growth. The lack of close relationship of temperature and economy is really a pity.

Stanton et al. (2009) divided IAMs into two categories: (1) the inter-temporal optimization model, this kind of model based on the principle of global or regional welfare maximum or minimum cost to get the optimal path of the future, the so-called "optimal" is the best hypothesis for the future the model can make and future state of the world modelers expected. (2) the simulation model, also known as the assessment model, this model is to evaluate different policies, not make assumptions or to seek the optimal in the future. These two kinds of model is nonlinear, need a lot of exogenous variables to express the economic and natural system in the model.

Due large categories of IAM, each has its own characteristics. some defects of one type of IAM, may cease to exist in another type of IAM (many IAM model can be regarded to be complementary on the function, can't simply say one model is better than the other one). In addition, an obvious characteristic is that more transparent model mechanism is, more vulnerable to criticism, such as RICE,

Table 1.1 Primary IAM model

Name	Global/regional	National economic level/sectoral economic level	Inter-regional economic association	Interdepartmental connection	Optimal/simulation	Loss evaluation of temperature rise	Temperature feedback on the economic system
AIM	Aisa-Pacific	National economic level, considering the energy supply and demand balance	–	N	Simulation	Y	N
IMAGE	Global	5 sectors	N	N	Simulation	Y	N
MESSAGE	Global	National economic level/	N	–	Optimal	Y	N
MARIA	Global	National economic level, considering the energy supply and demand balance	Y	N	Optimal	Y	Y
MiniCAM	Global	National economic level, considering the energy supply and demand balance, Agricultural supply and demand balance	N	N	Simulation	Y	Y
RICE	Global	National economic level/	N	–	Optimal	Y	Y
MRICES	Global	National economic level/	Y	–	Both	Y	Y
MERGE	Global	National economic level/	N	–	Optimal	Y	Y
WITCH	Global	National economic level/	N	–	Optimal	Y	Y
FUND	Global	Sectoral economic level	Y	Y	Simulation	Y	N
GREEN	Global	Sectoral economic level	Y	Y	Simulation	N	N
G-CUBED	Global	Sectoral economic level	Y	Y	Optimal	N	N
WIAGEM	Global	Sectoral economic level	Y	Y	Simulation	Y	N

MERGE. However, criticism is beneficial, it always provide directions for the model's development.

Goodess et al. (2003) pointed out that because IAM need to integrate different disciplines' content, so that may complicate the model. In order to simplify the process, a lot of IAM used several relatively simple equations to depict the corresponding mechanism, which is more noticeable in the climate system and the carbon cycle system, some IAM models with a few of equations to describe the climate system (Goodess et al. 2003). In fact, the IAMs he described is mainly focused on cost-benefit analysis of optimization model and some CGE model, such as RICE, DICE, MERGE, FUND, Wiagem. Influenced by algorithm, for the CGE, its climate module is simple, the relationship between radiation force and temperature is linear equation. This is subject to linear characteristics of CGE calculation equations. Due to the inability of using linear function to represent large dynamic climate models, the climate module of CGE could not be too complicated. For optimization model, when the climate module is too complex, it will cost a lot of computation time, even can not get the solution. However, there are some attempts to change this situation, a new algorithm OBOT (oracle based optimization technique) has made some breakthrough, firstly it decomposes IAM into two sub modules, climate and economic module, and then uses the database interaction search technology of the two sub-modules to complete the optimal path of the search (Bahn et al. 2006). This technology can extend climate module of RICE model in more detail, at the same time keep the important features of RICE, the original economic and climate module combined closely. A possible foresight is that the technology can expand RICE economy module for more specific CGE module. That may now face two difficulties, first, because the CGE has yet to include climate change factors into the production function; second, searching for the optimal strategy on department level will increase huge calculation amount.

1.3 IAM Modeling Principle

DICE model and RICE model are the most typical ones among various IAM models. We can understand the modeling principle of IAM easily through the understanding of them.

DICE model is the abbreviation of Dynamic Integrated model of Climate and Economy and RICE model is the abbreviation of Regional Integrated model of Climate an Economy, which is based on the development of DICE model. DICE model/RICE model are modeled and developed by some climate economist, leading by William Nordhaus, in Yale University. They established a series of economically dynamic process model, including DICE model (Nordhaus 1992), DICE-2007 model (Nordhaus 2007), RICE model (Nordhaus and Yang 1996), RICE-99 model (Nordhaus and Boyer 2000), RICE-2007 model (Nordhaus 2008) and so on.

DICE model published in 1992, a prototype of the work of Nordhaus in 1979 (Nordhaus 1992, 1994). DICE model integrates a general equilibrium model of global economy and a climate system that includes greenhouse gas emissions, carbon dioxide concentration, climate change, climate change impact and optimal policy. Therefore, DICE model is the IAM for optimizing policy (also known as welfare maximization model), whose behaviors of saving and investment are based on Ramsey model, and are developed by using GAM platform. The world is regarded as a whole in DICE model and countries and regions aren't distinguished. Comparing other IAM models during the same period, whose spatial scale can be reduced to regions even grids, such as FUND, AIM, IMAGE and so on, DICE model has a larger spatial scale. DICE model is more focused on the quantitative impact of climate change on economy and analysis of gains and losses in world economy owing to the implementation of climate protection policy (Goodess et al. 2003).

DICE model directly obtains the economic cost of climate change from the equation of climate damage (Goodess et al. 2003), in which the reaction of economies to climate change can be shown by investment change. The core of DICE model is policy instruments to control greenhouse gas emissions rate, which is a reduction ratio of global greenhouse gas emissions controlled by reduction rate to baseline scenario.

The climate system in DICE model is relatively simple comparing to other IAM of using GCM directly, for example FUND. It is a Simple Climate Mode (SCM) of Box-Advection Model and calculates annual average global temperature change with an interval of 10 years from 1965 to 2105. Meanwhile, DICE model uses computational results from annual average global temperature from 1862 and 1989 (Jones et al. 1990) and three GCM models (Schneider and Thompson 1981; Stouffer 1989; Schlesinger and Jiang 1990) to calibrate the computational results from simulation of climate model. In the processing of uncertainty, DICE model uses Monte Carlo, a way of random distribution dealing with uncertainty of parameters and adopts different climate sensitivity.

RICE model is modeled by Nordhaus and Yang 1996. Compared to DICE model, the biggest feature of RICE model is dividing the world into six regions that are China, the United States, Europe, Japan, the former Soviet Union and the rest parts of the world. The structure of equations in RICE model is basically consistent with DICE model and is developed by GAMS platform. From this perspective, earlier RICE model is equivalent to a multi-regional version of DICE model. RICE model uses the way of changing the intercept of fitting parameters to estimate different mitigation costs of each region, which are distinguished with DICE model. From the calculation results, RICE model gets much higher results than DICE model about world outputs and greenhouse gas emissions in the end of 21 century.

Nordhaus and Boyer (2000) developed a new version of the RICE model, namely RICE-99 model. RICE-99 model adopts a different modelling method with earlier version, in which the structure of model and control variables are changed and the model is depicted more sophisticated. In addition, RICE-99 model is developed by programming in EXCEL rather than in GAMS platform. The main differences between RICE-99 model and RICE model are: First, RICE-99 adopts a more complicated Cobb-Douglas production function (CD function) with three factors: capital, labor and energy, while DICE model and RICE model use CD function with two factors: capital and labor. RICE-99 model changes production relations of the whole model, namely that economic growth is a function of energy use, which means that the effect of reduction emissions is considered more in industrial processes. Secondly, energy supply is combined with fossil fuel consumption and the consumption of energy is decided by the market. Thirdly, RICE-99 model adopts a three-layer carbon cycle model, including atmospheric carbon flow, shallow and deep marine biosphere to replace a single system in original RICE model but still retain the original temperature dynamic mode. Fourth, RICE-99 model changes the impact of climate change on economy, making the global impact of climate change derive from regional impact.

Nordhaus and Boyer (2000) developed a new version of the RICE model, namely RICE-99 model. RICE-99 model adopts a different modelling method with earlier version, in which the structure of model and control variables are changed and the model is depicted more sophisticated. In addition, RICE-99 model is developed by programming in EXCEL rather than in GAMS platform. The main differences between RICE-99 model and RICE model are: First, RICE-99 adopts a more complicated Cobb-Douglas production function (CD function) with three factors: capital, labor and energy, while DICE model and RICE model use CD function with two factors: capital and labor. RICE-99 model changes production relations of the whole model, namely that economic growth is a function of energy use, which means that the effect of reduction emissions is considered more in industrial processes. Secondly, energy supply is combined with fossil fuel consumption and the consumption of energy is decided by the market. Thirdly, RICE-99 model adopts a three-layer carbon cycle model, including atmospheric carbon flow, shallow and deep marine biosphere to replace a single system in original RICE model but still retain the original temperature dynamic mode. Fourth, RICE-99 model changes the impact of climate change on economy, making the global impact of climate change derive from regional impact.

Nordhaus and Yang have begun to the development of the new version of RICE model since 2002 and finally developed RICE-2007. Compared to previous versions, regional division of RICE-2007 is more detail, and the model has shorter intervals and longer time span.

DICE model sees the world as an unity, thus it cannot make a distinction between different national emission models and emission-cutting policies. In order to analyze the role of national(regional) emission-cutting policies for the change of global climate, Nordhaus and Yang (1996) developed 6 national(regional) RICE (Regional Integrated model of Climate and the Economy) model. Hereafter, the model is extended in 8 national(regional), 12 national(regional) RICE models. From the perspective of the structure of model's equation, RICE is very similar to DICE. The primary distinction of both models is that the estimation of parameters are put into the regional levels, when RICE describes national(regional) economic behaviors and climate change. In other words, in RICE model, nations(regions) have independent economic behaviors and climate change exerts different effects on nations, but the change of global climate system's status is shared by nations. The structure of model is as follows:

- Object function

In the DICE model, intertemporal maximization of social welfare model serves as the objective function, as all consumer choice and emission-cutting policies are in the direction of evolution that is conducive to the object. Through the expressive method of optimal economic growth theory's utility, intertemporal social welfare is defined as the discounted value of per capita consumption, under the effect of weight of population scale. Formula (1.1) is its computational equation.

$$W = \sum_{t=1}^{T\max} U[c(t), L(t)]R(t) \tag{1.1}$$

where W is social welfare, c is per capita consumption, L is population, and R is the rate of discounted value. Further, $U[]$ can be defined as:

$$U[c(t), L(t)] = L(t)\left[c(t)^{1-\alpha}/(1-\alpha)\right] \tag{1.2}$$

where α is constant elasticity of the marginal utility of consumption, which describes alternative between different generations. When the value of α is zero, consumptions between different generations can be replaced on a large measure; when α is larger, consumptions between different generations aren't increasingly replaced. In addition, the discounted value R in formula (1.1) can be estimated by formulated (1.3).

$$R(t) = (1+\rho)^{-t} \tag{1.3}$$

where ρ is the preference of social time, which gives different weights to different generations by formula (1.3). When determining value of ρ is smaller, the future utility is more important; when the value of ρ is zero, utilities between different generations have identical importance. With reference to the issue of value of discounted value, it is the front-burner issue of present climate protection modelling. In its essence, determining value of ρ involves the issue of climate protection ethics, as is the same as the issue that whether future person's consumptions are as important as present person's consumptions, debated by Nordhaus and Stern. Nordhaus (2007) advocated that the value of ρ should be determined with 0.015 by the estimation of practical experience, while Stern, holding the contradictive view that the future consumption is as important as present consumption. argued the value of ρ should be 0.001. The value is very close to zero, which lets future utility fully be discounted, thus the estimation of impact of climate change on future welfare may be over-estimated.

However, the objective functions of RICE model and DICE model are a little different, which can be seen in (1.1')

$$W = \sum_{t=1}^{T\max} \sum_{i=1}^{N} \psi_{I,t} U^{(i)} \left[C^{(i)}(t), L^{(i)}(t) \right] R^{(i)}(t) \tag{1.1'}$$

Where, $\psi_{I,t}$ denotes the weights of countries or regions. In other words, the objective function of RICE is the sum of countries' or regions' utilities which is calculated by weights. The advantage of this objective function is that more scenarios can be considered by adjusting the weights.

- Economic System

DICE/RICE model has the same production function as (1.4), which adopts the C-D production function with the feature of constant returns to scale.

$$Q(t) = [1 - \Lambda(t)]A(t)K(t)^{\gamma}L(t)^{1-\gamma} / [1 + \Omega(t)] \tag{1.4}$$

Where, $Q(t)$ is the net output, $A(t)$ is the total factor productivity, $K(t)$ is the capital, capital is accumulated by the perpetual inventory method, satisfying:

$$K(t) = I(t) + (1 - \delta_K)K(t - 1) \tag{1.5}$$

Where, $I(t)$ is investment, δ_K is the discount rate; Labor force grows at a rate decreasing gradually. In addition, $\Omega(t)$ and $\Lambda(t)$ is the economic loss rate caused by climate change and the abatement cost ratio, satisfying:

$$\Omega(t) = 1/\{\psi_1 T_{AT}(t) + \psi_2[T_{AT}(t)]^2\} \tag{1.6}$$

$$\Lambda(t) = \Psi(t)\theta_1(t)\mu(t)^{\theta_2} \tag{1.7}$$

Where T_{AT} represents the temperature rise, which is written as T for short; $\mu(t)$ is the abatement rate, ψ_1, ψ_2, θ_1, θ_2, $\Psi(t)$ are the parameters. Nordhuas sets

$$\Omega(t) = 1/(1 + D[T(t)/3]^2) \tag{1.8}$$

And according to experience data, we set $D = 0.0133$

$$\Lambda(t) = b_1\mu(t)^{b_2} = 0.0686\mu(t)^{2.887} \tag{1.9}$$

$$\Omega = [1 - b_1\mu(t)^{b_2}]/[1 + d(t)] = [1 - 0.0686\mu(t)^{2.887}]/[1 + 0.00144T(t)^2] \tag{1.10}$$

In fact, the model is equivalent to modifying the total factor productivity:

$$A^* = A/[1 + D(T(t)^2/9)] \quad (*)$$

Function (*) has the obvious economics significance. A^*, represents the climate change influenced the total factor productivity loss of output, and it is called the effective productivity. If further consideration reduction activity, let

$$A^* = A(1 - \Lambda(t;\mu))/[1 + D(T(t)^2/9)] \quad (**)$$

Furthermore, Nordhaus and Yang (1996) considered the total output productivity A_i is different across countries. Thus,

$$\ln A_i(t) = \ln A_i(t-1) + \gamma_{i,a}\exp(-\delta_{i,a}t) + \sigma_{i,a}\varepsilon_t \quad (1.11)$$

Where $\gamma_{i,a}$,$\delta_{i,a}$,σ_a are parameters. And ε_t is the standard Independent identically distributed random disturbance on the other hand, the output is used for consumption $C_i(t)$ and investment $I_i(t)$.

$$Q_i(t) = C_i(t) + I_i(t) \quad (1.12)$$

In DICE/RICE model, economic output yields the carbon emission. Industrial carbon emission E_{Ind} is calculated by the economic output and the intensity of carbon emission.

$$E_i^{\text{ind}}(t) = \sigma_i(t)[1 - \mu_i(t)]A_i(t)K_i^{\gamma}(t)L_i^{1-\lambda}(t) \quad (1.13)$$

The ith country's carbon emission intensity is exogenous. The accumulated carbon emission caused by fossil fuel is constrained by

$$C_{Cum} \geq \sum_{t=1}^{T\max} E_i^{Ind}(t) \quad (1.14)$$

Besides the industrial carbon emission, the change of land use is also an important carbon emission source. It is estimated that land use change can cause nearly 1.5 GtC carbon emission. Thus, the total carbon emission $E(t)$ is the sum of the industrial emission $E^{Ind}(t)$ and the emission of land use change $E_{Land}(t)$.

$$E(t) = E^{Ind}(t) + E_{Land}(t) \quad (1.15)$$

Where, the emission of land use change $E_{Land}(t)$ is exogenous.

- the geophysical system

Carbon emissions from economic activity will affect the land, ocean and the atmospheric carbon concentration, leading to global radiation force change, finally result in global warming. DICE/RICE model links the geophysical system and economic activity as follow:

$$M_{AT}(t) = E(t) + \phi_{11}M_{AT}(t-1) + \phi_{21}M_{UP}(t-1) \tag{1.16}$$

$$M_{UP}(t) = \phi_{12}M_{AT}(t-1) + \phi_{22}M_{UP}(t-1) + \phi_{32}M_{LO}(t-1) \tag{1.17}$$

$$M_{LO}(t) = \phi_{23}M_{UP}(t-1) + \phi_{33}(t-1) \tag{1.18}$$

Where, $M_{AT}(t)$ is the atmospheric carbon inventory $M_{UP}(t)$ is the up ocean and biology cycle carbon inventory and $M_{LO}(t)$ is carbon inventory of deep ocean. ϕ_{ij} represents the carbon transforming coefficients. Thus the global radiation force level $F(t)$:

$$F(t) = \eta\{\log_2[M_{AT}(t)/M_{AT}(1750)]\} + F_{EX}(t) \tag{1.19}$$

Where $M_{AT}(1750)$ represents the carbon concentration before year 1750. Because DICE/RICE model is mainly used for assess the carbon emission effect, other greenhouse gases such as methane effect of nitrous oxide, etc. are not included. And they are denoted as $F_{EX}(t)$ in the model:

$$T_{AT}(t) = T_{AT}(t-1) + \xi_1\{F(t) - \xi_2 T_{AT}(t-1) - \xi_3[T_{AT}(t-1) - T_{LO}(t-1)]\} \tag{1.20}$$

$$T_{LO}(t) = T_{LO}(t-1) + \xi_4\{T_{AT}(t-1) - T_{LO}(t-1)\} \tag{1.21}$$

Where $T_{AT}(t)$ is the earth surface temperature, $T_{LO}(t)$ is the deep ocean temperature.

1.4 Global Carbon Cycle Model

Since the DICE/RICE model was developed, not only Nordhaus the first person proposed this model continue to improve this model, have launched a new version of the model, and other scholars based on the DICE/RICE have made a lot of improvements.

For example, Bosetti et al. (2006) introduced the endogenous technological progress caused by the interaction between R & D investment and learning by doing into the model; Wang et al. (2010) based on the DICE/RICE model built a China-US climate protection model including GDP spillover mechanism; Zwaan et al. (2002) developed the DEMETER model based on the DICE/RICE model by introducing learning by doing mechanism within the energy system; Popp (2004a, b) also developed the ENTICE model based on the DICE/RICE model; But the improvement of the scholars, mainly concentrated in the economic system, there is little of the model of the earth's physical system.

In theory, we can achieve a detailed description of the atmospheric motion process through the thermodynamics and fluid mechanics equations in the GCM model, which is one of the ideal methods to establish the model of climate protection. However, because the GCM model involves thousands of equations, it is not well integrated into the policy optimization model of climate protection. Instead, the approach is based on the assumption that the global carbon cycle is characterized by a much simpler model of the earth's physical system, combined with the modeling of climate protection.

At present, the simplified carbon cycle models can be divided into two types: one is the whole model or the zero dimensional model, such models describe the global carbon cycle as a whole by describing the carbon cycle of the atmosphere, the earth's biosphere, the soil, the oceans, and so on; the second model is the spatial distribution model and the model based on geo spatial distribution as the background, taking into account the different regions with different geographical conditions, vegetation type, soil type factors, which constitute the refinement of the global carbon cycle, and then analyze the effect of different regions in the global climate change by the difference. By comparison, the former model is easy to implement, and the latter puts forward higher requirements to the model data, which is difficult to achieve.

Now, we will introduce a zero dimensional model of the earth physics system, which can be integrated with the DICE/RICE model well.

- The climate model:

Let $T(t)$ be the global surface temperature.

$$\frac{dT(t)}{dt} = \mu \ln \frac{C(t)}{C_0} - \alpha T(t) \tag{1.22}$$

Where $C(t)$ is the total amount of carbon in the atmosphere, C_0 is the carbon content in the atmosphere before industrialization, μ and α are the model parameters, which are 0.17 and 0.034, there is the function relationship between surface temperature and atmospheric carbon content and the temperature change.

- Terrestrial carbon cycle

Terrestrial carbon is shared between two compartments: biota (vegetation) and pedosphere (soils). Let us denote $N(t)$ as the amount of carbon in vegetation.

$$\frac{dN(t)}{dt} = P(C, N, T) - m(t)N(t) \tag{1.23}$$

Where $P(C, N, T)$ is the annual productivity or the net primary, measured in Gt/yr. $m(t)$ is the carbon escape rate in vegetation.

$$P(C,T) = P_0(1 + a_1 T)(1 + a_2(C - C_0)) \tag{1.24}$$

Where P_0 is the pre-industrial value of NPP.

$m(t)$ is defined as:

$$m(t) = \frac{1}{\tau_B(t)} \tag{1.25}$$

Where $\tau_B(t)$ is the residence time of carbon in vegetation. That is to say, the escape rate of vegetation carbon is inversely proportional to its retention time. Carbon escape from the vegetation is divided into long-term and short-term retention of two types; the former will be converted to soil carbon, which will be released in the form of carbon dioxide into the atmosphere.

We suggest that ε indicates that the proportion of the long-term carbon emission from the biomass, and therefore the proportion of short-term carbon is $1-\varepsilon$. Finally, the dynamics for the amount of carbon in soils is written as:

$$\frac{dS(t)}{dt} = \varepsilon m(t)N(t) - \delta(T)S(t) \tag{1.26}$$

$\delta(T)$ is the decomposition rate of soil carbon. In addition to the amount of soil carbon changes in the amount of carbon released from the vegetation, but also through the degradation process to release some of the carbon.

$$\delta(T) = \delta_0(1 + a_3 T) \tag{1.27}$$

- Ocean carbon cycle

Each period, marine carbon content changes not only by marine influence on a carbon concentration, and also affected by the atmospheric carbon levels, namely the carbon in the atmosphere will be into the ocean circulation.

$$\frac{dD(t)}{dt} = Q_{oc} = \sigma[(C(t) - C_0) - v(D(t) - D_0)] \tag{1.28}$$

Where σ, ξ are the model parameters

- Atmosphere carbon cycle

Under the action of terrestrial carbon cycle and ocean carbon cycle, carbon fluxes in the atmosphere is below.

$$\frac{dC(t)}{dt} = -P(C,T) + (1 - \varepsilon)m(t)N(t) + \delta(T)S(t) - Q_{oc} + E(t) \tag{1.29}$$

Where $E(t)$ is the carbon emissions from human activities.

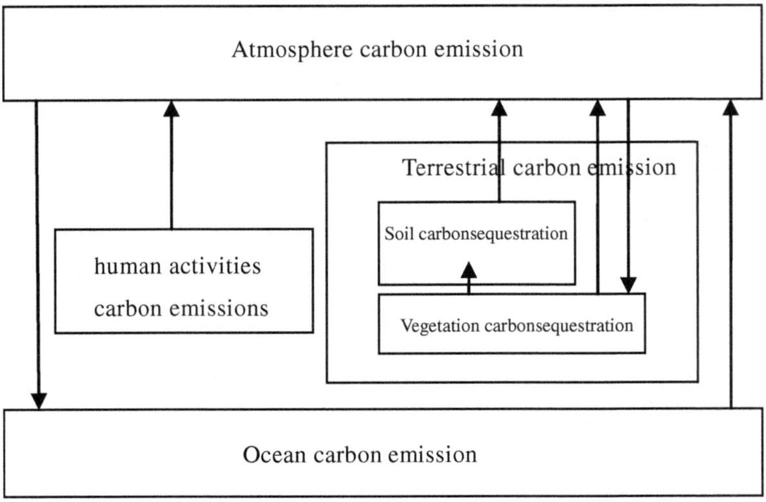

Fig. 1.1 Global carbon cycle

To sum up, the global carbon cycle model built by Svirezhev et al. (1999) is a three layer zero dimensional model based on the atmosphere, land and ocean carbon pool as the main body (Fig. 1.1).

1.5 Shortcomings

Although DICE, RICE series model continue to improve themselves in the long-term of the development and application, there are still some problems. DICE model has no regional division and takes the world as a whole. Therefore, there is only one total utility function in the world. Thus the marginal utility of income in this model can only be compared across periods and can not be compared among regions. So it is unable to analyze the global or the regional climate adaptation policies. The climate model of DICE model is relatively simple and its practicability is limited. In addition, there is no point in taking the extreme climate events into consideration in the global mean climate system. For this reason, the global damage equation of extreme climate events needs to be constructed on the basis of the regional extreme index. For example, DICE and CETA model of such a global scale IAM cannot establish the global extreme or sudden climate events of the damage equation (Goodess et al. 2003).

Although RICE model divides the world into a number of regions, there is no economic link among the regions. And this is obviously unreasonable as the relationship among the current countries economy in the world are more and more close. In addition, the RICE model does not take into account the endogenous progress of technology and thus we cannot estimate the economic development of developing countries to reduce the speed of energy use. Therefore, there is a big gap in the forecast of the world economic development and carbon emission trends, when comparing with the real situation of the world economic development since 1996.

Limited to the characterization of the economic system, DICE model series defines the energy intensity decline rate to simulate the decline in the use of energy per unit of GDP in the process of economic development.

A shortage of popular DICE or RICE is that they are written by GAMS or EXCEL. This affects the calculation speed or the technology accuracy.

A criticism of the general IAM is presented by Wang Zheng. Wang Zheng believes that the economy of the various regions of the RICE is always related to each other (Wang et al. 2009). Wang et al. adopt the method of Douven and Peeters (1998). The RICE model was improved by adding the Mundell-Fleming model to connect each region economically. However, the linear mechanism of Mundell-Fleming model does not fully reflect the complexity of the world economic integration. So, we can introduce complete mechanism to the RICE to improve its lack of regional economic links. The introduction of the Mundell-Fleming model in the RICE produce of MRICES (Wang et al. 2008).

Another criticism is that the economic system is relatively simple. Ackerman et al. (2009) believed that IAM, represented by RICE and DICE, has two important defects—the lack of considering the equilibrium of the market and the lack of no internal biochemistry of DICE's technological progress. In view of these two kinds of situations, Wang and Wu et al. (2010) has been the introduced the Learning by doing mechanism into the RICE to produce the MRICES. Wang and Gu (2016) introduced the global market equilibrium mechanism into the MRICES to produce the CIECIA. The narrative constitutes the main content of this book.

From the existing IAM model, climate change as a factor affecting economic development is introduced into the production function based on the optimal MERGE and RICE models. However, it still has some problems in economic refinement, technological progress and regional economic relations. And large scale integration evaluation model including CGE type model, such as ICAM, IMAGE, FUND, WIAGEM, cannot incorporate the climate change factor into the economic growth model. So, establishing a new IAM is imminent. This new IAM will starts from the theory of economic growth, and fully considers the economic growth model under the mode of climate change. Specifically, in the context of climate change, the factors that affect economic growth are capital, labor, technology, climate change, industrial structure, energy structure and so on. Climate change factor refers to the carbon concentration of Nordhaus (2008) and temperature as a "negative impact" of the natural capital affects the productivity.

The adjustment of industrial structure will also leads to the changes of energy intensity, so the industrial structure will bring an impact on economic growth. Moreover, for specific countries or regions, the implementation of emission reduction programs to the industry is also a realistic problem. "In 2007, at the conference of the United Nations, 13 session of the conference of the parties, the Bali Action Plan (BAP) requires developing countries to take "measurable, reporting and verification (MRV)" of the appropriate mitigation actions (NAMAs) to reduce greenhouse gas emissions, which needs the capital and technology of

MRV and the capacity building support in developed countries (Ott et al. 2008)". This requirement makes it an inevitable trend to study carbon emissions from the industrial level.

Another problem is that the department reducing emission tasks can not be arbitrarily assigned and need to take into account the industrial balance. The input-output theory Leon Leontief tells us there are complex staggered relationships among industries. The change of output and input in one sector can be spread to many other sectors. In order to show the industrial structure and the energy structure into the models, it will also make a more detailed level of emission reduction plan for the future. Usually, this means the new IAM model will maintain the mechanism of climate impact on economic growth and extend to the CGE model. And then improving the traditional dynamic CGE and increasing the mechanism to change the trend of industrial structure evolution can assess the impact of industrial structure adjustment on the economy and global climate. This book describes the MRICES-2014 to develop such a balanced economy of IAM. Unfortunately, limited to China's communications conditions, this work only achieves in the Chinese economy.

In addition, there is a fundamental lack of IAM. It is not reasonable in ethics to emphaze the global emission reduction benefit optimization or global welfare maximum. Global emission reduction benefit optimization may mean that the developed countries get more emission rigdue to advanced technology and finance, capital advantage position and also mean that the opportunity of developing in the developing countries is deprived. For this problem, Wang and Wu (2010) abandon the global emission reduction benefits of the optimal target in their MRICES model and horizontally compare to try to circumvent the ethical problems. Yang and Sirianni (2010) proposed the non color changing principle and take it as a coordination standard to solve this ethical problems.

References

Ackerman F, DeCanio SJ, Howarth RB, Sheeran K (2009) Limitations of integrated assessment models of climate change[J]. Clim Change 95(3–4):297–315

Bahn O, Drouet L, Edwards NR et al (2006) The coupling of optimal economic growth and climate dynamics[J]. Clim Change 79(1–2):103–119

Bosetti V, Carraro C, Galeotti M et al (2006) WITCH: a world induced technical change hybrid model. Energy J Spec Issue Hybrid Model Energy Environ Polic Reconcil Bottom-up Top-down:13–38

Bruckner T, Hooss G, Füssel HM, Hasselmann K (2003) Climate system modeling in the framework of the tolerable windows approach: the ICLIPS climate model[J]. Clim Change 56 (1–2):119–137

Douven R, Peeters M (1998) GDP-spillovers in multi-country models[J]. Econ Model 15(2): 163–195

Goodess C, Hanson C, Hulme M, Osborn T (2003) Representing climate and extreme weather events in integrated assessment models: a review of existing methods and options for development. Integr Assess 4:145–171

Jones PD, Wigley TML, Wright PB (1990) Global and hemispheric annual temperature variations between 1861 and 1988. Carbon Dioxide Information Centre, NDP-022 = R1. Oak Ridge National Laboratory, 1990

Kemfert, C (2002) An integrated assessment model of economy-energy-climate-the model Wiagem[J]. Integr Assess 3(4):281–298

Lucas, RE (2002) Lectures on economic growth[M], Harvard Univ Pr

Nordhaus WD (1979) Efficient use of energy resources. Yale University Press, New Haven, CT

Nordhaus WD (1992) An optimal transition path for controlling greenhouse gases. Science 258 (5086):1315–1319

Nordhaus WD (1994) Managing the global commons: the economics of climate change. MIT Press, Cambridge, Mass

Nordhaus WD (2007) Accompanying notes and documentation on development of DICE-2007 model: notes on DICE-2007.v8 of September 28

Nordhaus WD (2008) A question of balance: weighing the options on global warming policies. Yale University Press

Nordhaus WD, Boyer J (2000) Warming the world: economic models of global warming. The MIT Press, Cambridge, Massachusetts London, England

Nordhaus WD, Yang Z (1996) Regional dynamic general-equilibrium model of alternative climate-change strategies. Am Econ Rev 86:741–746

Popp D (2004a) ENTICE: endogenous backstop technology in the DICE model of global warming. J Environ Econ Manag 48(1):742–768

Popp D (2004b) ENTICE: endogenous backstop technology in the DICE model of global warming. J Environ Econ Manag 2004(48):742–768

Rotmans J, De Boois H, Swart RJ (1990) An integrated model for the assessment of the greenhouse effect: The Dutch approach[J]. Clim Change 16(3):331–356

Schlesinger ME, Jiang XJ (1990) Simple-model representation of atmosphere-ocean GCMs and estimation of the timescale of CO2-induced climate change. J. Clim 3:1297–1315

Schneider SH (1997) Integrated assessment modeling of global climate change: Transparent rational tool for policy making or opaque screen hiding value-laden assumptions?[J]. Environ Model Assess 2(4):229–249

Schneider SH, Thompson SL (1981) Atmospheric CO2 and climate: importance of the transient response. J Geophys Res 86(1981):3135–3147

Stouffer RJ, Manabe S, Bryan K (1989) Interhemispheric asymmetry in climate response to a gradual increase of atmospheric CO2. Nature 342:660–662

Svirezhev Y, Brovkin V, Bloh W, Schellnhuber HJ, Petschel-Held G (1999) Optimisation of reduction of global CO2 emission based on a simple model of the carbon cycle. Environ Model Assess 4:23–33

Tol RS (2002) Welfare specifications and optimal control of climate change: an application of fund [J]. Energy Econ 24(4):367–376

van der Zwaan BCC, Gerlagh R, Klaassen G et al (2002) Endogenous technological change in climate change modeling. Energy Econ 24:1–19

van Vuuren DP, Lowe J, Stehfest E et al (2011) How well do integrated assessment models simulate climate change?[J]. Clim Change 104(2):255–285

Wang Z, Li HQ, Wu J et al (2010) Policy modeling on the GDP spillovers of carbon abatement policies between China and the United States. Econ Model 27(1):40–45

Weyant J, Davidson O, Dowlabathi H et al (1996) Integrated assessment of climate change: an overview and comparison of approaches and results[M]. Cambridge University Press, Cambridge, United Kingdom and New York, NY, USA

Yang Z (2008) Strategic Bargaining and Cooperation in Greenhouse Gas Mitigations: An Integrated Assessment Modeling Approach[M], MIT Press

Yang Z, Sirianni P (2010) Balancing contemporary fairness and historical justice: A 'quasi-equitable'proposal for GHG mitigations[J]. Energy Econ 32(5):1121–1130

Chapter 2
MRICES

2.1 Introduction

Since the 1990s, climate change policy simulation has been developed in two categories, the Integrated Assessment Models (IAM) and the Computable General Equilibrium (CGE) model. The climate change IAM is recognized as a broad, integrated model that analyzes scientific and socioeconomic aspects of climate change primarily for the purpose of assessing policy options for climate change control (Kelly and Kolstad 1999). IAM research includes the DICE and RICE models established by Nordhaus (1991) and Nordhaus and Boyer (2000), respectively; the FOUND model established by Link and Tol (2004); the MERGE model established by Manne et al. (1995) and Manne and Richels (2004); and the TIAM and ETSAP-TIAM models established by McGill (Loulou and Labriet 2007; Loulou 2008). Models based on the IAM in China have also been developed by Jiang and Hu (2002), Chen et al. (2004), Wang et al. (2002, 2004), among others. For reference, CGE-based models of climate policy simulation have been developed by Dixon et al. (1982), the OECD (1992), Burniaux et al. (1992), Mckibbin and Wilcoxen (1993), Nijkamp et al. (2005) and Bussolo et al. (2008).

In this chapter, we focused on an IAM-based model. The most representative IAM-based models of climate change are the DICE (Dynamic Integrated model of Climate and the Economy) and RICE (Regional Integrated model for Climate and the Economy) model. In 1990, Nordhaus first presented the DICE model (Nordhaus 1990), which contains no regional disaggregation, so that policymakers cannot implement region-specific policies. As an extension of DICE, RICE is a multiregional model that divides the world into six regions: the United States, China, the European Union, Japan, the former Soviet Union and the rest of the world (Nordhaus and Yang 1996). In 1999, a new version of RICE separated the world into eight groups: the United States, other high income countries, OECD Europe, Russia and Eastern Europe, middle income countries, lower-middle income countries and low income countries (Nordhaus and Boyer 2000). The more detailed a

Z. Wang et al., *Integrated Assessment Models of Climate Change Economics*, DOI 10.1007/978-981-10-3945-4_2

region is modeled, the more accurate the results are produced for a specified regional policy simulation. There are many multiregional models of climate policy, including the 25-region WIAGEM model (Kemfert 2005), the 6-region ETC-RICE model (Buonanno et al. 2003) and the 32-country POLES model, among others. The DICE and RICE models as a paradigm for climate change modeling have been undergoing continuous improvements; however, there are still rooms for further modifications. In this study, we introduced multilateral GDP spillovers into the IAM.

Although spillovers in climate change modeling have been studied for a long time, most researches have focused on technological spillovers (Rao et al. 2006; Golombek and Hoel 2003) and R&D spillovers (Buonanno et al. 2003) that would affect mitigation technology. However, modern countries have become much closer through trade and cooperation than they have been at any time in the past. GDP growth causes international spillovers; i.e., GDP growth in one country may have a positive or negative influence on other countries. For example, Luh and Shih (2006) discovered that spillover contributed to GDP growth between Japan and Korea but dampened Taiwan's GDP growth between 1978 and 1992. In particular, the 2008 global financial crisis, which was initially triggered by the bursting of the U.S. housing bubble, spread to all regions of the world, resulting in dire consequences for global trade, investment and growth. With the 2008 financial crisis, India's 2008–09 GDP growth decreased to approximately 7.1% from 9% annual growth before (Jha 2009). Weeks (2009) reported that the 2008 global financial crisis caused a fall in export earnings in Sierra Leone of approximately 15% in 2009 compared to 2008. Liu (2009) found that the impact of the global financial crisis on China's economy was sizeable; a 1% decline in economic growth in the US, the EU and Japan is likely to result in a 0.73% decline in growth in China. International economic interactions have become so strong and complex that no country can remove itself from the global economy. Therefore, climate policies enacted in one nation will have an indirect economic influence on other countries via trade.

GDP spillovers in climate change can no longer be neglected; however, the number of researchers studying modeling of this issue is still small. Grubb et al. (2002) have divided the international spillover from GHG emissions reductions on developing countries into three aspects: spillovers from economic substitution, spillovers from diffuse technological changes and political effects on developing countries from the mitigation actions of industrialized countries. GDP spillovers from the mitigation actions of developed countries will significantly contribute to global emissions reduction (Grubb et al. 2002). Wang et al. (2010) combined the Mundell-Fleming model with a climate change model to examine the GDP spillover from climate policies between the United States and China. A crucial conclusion of Wang et al. (2010) was that one country should consider the emissions mitigation policies of other countries when enacting its own. Thus, the inclusion of GDP spillovers in models against climate change is essential; otherwise, assessments on climate policy are biased.

This study established the MRICES (Multi-regional integrated model of climate and economy with GDP spillovers) model, which integrates multilateral GDP spillovers into climate change modeling. Furthermore, several global mitigation strategies were assessed using the MRICES model to determine whether these strategies are climatically effective and economically efficient.

2.2 Model Description

MRICES is a dynamic model that combines climate and economic factors. It divides the world into six regions: the US, Japan, the European Union (EU), China, the former Soviet Union (FSU) and the rest of the world (ROW). Each region has its own economic system and shares the global climate system. The economic system describes the economic activities of each region, and global emissions can be determined on the basis of these activities. Global emissions will result in a temperature change in the global climate system, and the increase in temperature will then affect the economic system. The dynamic interaction between climate and economic systems is illustrated in Fig. 2.1.

2.2.1 *Economic System*

Based on the partitioning of the world in MRICES, the global economic system is also composed of six interdependent subsystems with identical modeling structures. The specific equations in the economic system are described below.

$$Y_{i,t} = A_{i,t}^{*} K_{i,t}^{\alpha} L_{i,t}^{1-\alpha} \tag{2.1}$$

Equation (2.1) describes the regional output based on the Cobb-Douglas production function from capital, $K_{i,t}$ and labor, $L_{i,t}$; i represents the number of the region, and t represents the simulation time, $A_{i,t}^{*}$ is a measure of exogenous productivity, and α and $1-\alpha$ represent the elasticity of capital and labor, respectively. Changes in $A_{i,t}^{*}$ depend on the exogenous productivity, $A_{i,t}$, according to Eq. (2.2), which has been

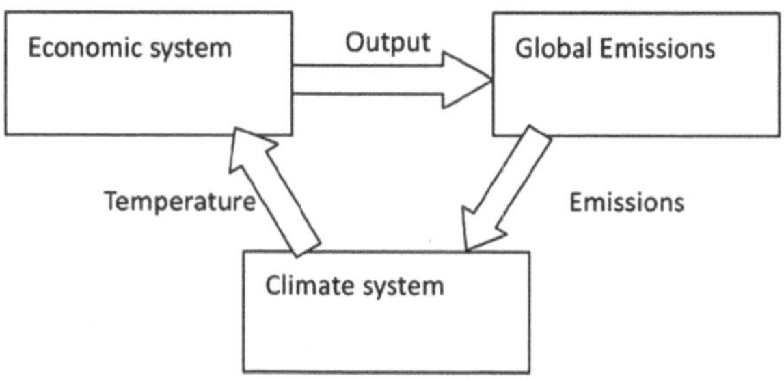

Fig. 2.1 The relationship between economic and climate systems in MRICES

adopted by Nordhaus and Yang (1996), Pizer (1999) and Eyckmans and Tulkens (2003), and $A_{i,t}$ grows with the annual decline in growth rate shown in Eq. (2.3).

$$A_{i,t}^{*} = \left(\frac{1 - b_{i,1}\mu_{i,p}}{1 + (D_0/9)T_{i,t}^{2}} \right) A_{i,t} \tag{2.2}$$

$$\ln(A_{i,t}) = \ln(A_{i,t-1}) + \gamma_{i,a}\exp(-\delta_{i,a}t) + \sigma_a \tag{2.3}$$

In these equations, $\mu_{i,p}$ describes the emissions reduction rate from production control, $b_{i,1}$ is the parameter for emissions control, D_0 is the fractional loss in aggregate GDP from a 3 °C temperature increase, $T_{i,t}$ is the average surface temperature (in °C) relative to that before industrialization, $\gamma_{i,a}$ is the initial growth rate of productivity, $\delta_{i,a}$ describes the annual decline in growth rate, and σ_a is the standard error for random shock.

In Eq. (2.1), the capital stock $K_{i,t}$ grows as

$$K_{i,t+1} + K_{i,t}(\delta_{i,K}) + I_{i,t}, \tag{2.4}$$

where $\delta_{i,K}$ is the capital depreciation rate, and $I_{i,t}$ denotes the investment during the corresponding period derived from

$$I_{i,t} = \eta_i Y_{i,t} - \eta_{i,I} - I_{i,t}^{(cs)} - I_{i,t}^{(f)} + I_{i,t}^{(n)} - M_{i,t}^{(f)} + M_{i,t}^{(n)}), \tag{2.5}$$

where η_i is the annual rate of investment, $\eta_{i,I}$ is the proportion of climate protection deducted from investment as a policy parameter, $I_{i,t}^{(cs)}$ is the investment in increasing carbon sinks, $I_{i,t}^{(f)}$ and $I_{i,t}^{(n)}$ are the investments in fossil and non-fossil fuel energy sources, respectively, and $M_{i,t}^{(f)}$ and $M_{i,t}^{(n)}$ are the maintenance costs of fossil and non-fossil fuel energy sources, respectively. The sum of the terms in parentheses in Eq. (2.5) is considered to be the cost of handling climate change. To determine the annual investment, household consumption can be expressed as

$$C_{i,t} = Y_{i,t} - I_{i,t} - (1 - \eta_{i,I})(I_{i,t}^{(cs)} - I_{i,t}^{(f)} + I_{i,t}^{(n)} - M_{i,t}^{(f)} + M_{i,t}^{(n)}). \tag{2.6}$$

Equations (2.5) and (2.6) implicitly indicate that two emissions reduction actions, carbon sinks and energy substitution need financial support. The source of emissions reduction costs is different from that of van der Zwaan et al. (2002), in which the cost for handling climate change is defined as part of output. In our model, the cost of emissions reduction is partially derived from consumption, with a rate of $(1 - \eta_{i,I})$ and partially derived from investment, with a rate $\eta_{i,I}$. As a result, actions against climate change influence the utility of each region, as shown by the Ramsey function:

$$U_i(n) = \sum_{t=1}^{n} (1+\rho_i)^{-t} L_{i,t}^{(s)} \frac{(C_{i,t}/L_{i,t}^{(s)})^{1-\tau_i}}{1-\tau_i}, \tag{2.7}$$

where $U_i(n)$ is the utility when take a control policy in region i, and ρ, τ describes the discount rate and risk aversion to consumption per capita across time.

2.2.2 Emissions Mitigation

As previously mentioned, the link between the economic and climate systems is the global emissions level, which is the sum of the six regional emissions. It can be determined from the following equations:

$$E_{i,t}^{(c)} = \sigma_{i,t}(1-\mu_{i,c})Y_{i,t}\left(\frac{A_{i,t}}{A_{i,t}^{*}}\right) \tag{2.8}$$

and

$$E_t = \sum_{i=1}^{6} E_{i,t}^{(c)}, \tag{2.9}$$

where $E_{i,t}$ is the regional emissions level, $\mu_{i,c}$ is the integrated reduction rate, E_t is the global emissions level, and $\sigma_{i,t}$ is an exogenous trend in emissions per output, which declines as

$$\ln(\sigma_{i,t}) = \ln(\sigma_{i,t-1}) - \gamma_{i,\sigma}\exp(-\delta_{i,\sigma}t) \tag{2.10}$$

where $\gamma_{i,\sigma}$ is the initial growth rate for the emissions trend, and $\delta_{i,\sigma}$ is the annual decline in the growth rate.

For emission mitigation, there are three reduction measures considered in MRICES model, including production controlling, energy substitution and carbon sinks increasing. The gross emission reduction rate $\mu_{i,c}$ is presented in Eq. (2.11).

$$\mu_{i,c} = \mu_{i,p} + \mu_{i,s} + \mu_{i,d} \tag{2.11}$$

where $\mu_{i,p}$, $\mu_{i,s}$, $\mu_{i,d}$ denote reduction rate of production controlling, carbon sinks increasing and energy substitution respectively. The cost for carbon sink increasing can be written as

$$I_{i,t}^{(cs)} = C_i^{(s)} \mu_{i,s} \sigma_{i,t} Y_{i,t} \left(\frac{A_{i,t}}{A_{i,t}^*} \right) \tag{2.12}$$

where $C_i^{(s)}$ is cost for increasing carbon sinks in a ton of emission reduction, $\sigma_{i,t}$ is the carbon intensity representing the trend of emission/output, which changes according to Eq. (2.10).

In order to calculate the use of fossil fuel and non-fossil fuel annually under the policy of energy substitution, the total energy consumption in each period has to be figured out as

$$En_{i,t} = \varsigma_{i,t} Y_{i,t} \left(\frac{A_{i,t}}{A_{i,t}^*} \right) \tag{2.13}$$

where $\varsigma_{i,t}$ is the exogenous trend for energy intensity per output which grows according to function (2.13).

$$\ln(\varsigma_{i,t}) = \ln(\varsigma_{i,t-1}) + \gamma_{i,\sigma} \exp(-\delta_{i,a} t) \tag{2.14}$$

Therefore, the amount of non-fossil fuel is proportionally related to gross energy consumption and the emission reduction rate of energy substitution.

$$N_{i,t} = F_{i,t} / \lambda_{i,,FN} (1 - \mu_{i,d}) \tag{2.15}$$

where $\lambda_{i,FN}$ represents the proportion of fossil fuel in total energy consumption. On the other hand, if there is no policy of energy substitution, Eq. (2.15) can be replaced as

$$N_{i,t}^* = F_{i,t}^* / \lambda_{i,FN} \tag{2.16}$$

where $N_{i,t}^*$, $F_{i,t}^*$ is the non-fossil and fossil consumption respectively if no energy substitution policy is applied. Since consumptions of fossil and non-fossil fuel make up the total energy use, we can further get the consumption of fossil fuel with Eq. (2.17), no matter energy substitution policy is implemented or not.

$$En_{i,t} = F_{i,t} + N_{i,t} \tag{2.17}$$

Thus, the consumption difference of fossil fuel or non-fossil fuel between with and without energy substitution policy is caught as

$$\Delta N_{i,t} = N_{i,t} - N_{i,t}^* \tag{2.18}$$

$$\Delta F_{i,t} = F_{i,t}^* - F_{i,t} \tag{2.19}$$

Ground on the energy consumption in each period, we introduce the learning-by-doing mechanism which is proposed by van der Zwaan et al. (2002) into our model.

$$I_{i,t-1}^{(n)} = g(X_{i,t}^{(n)})a_{i,N}\Delta N_{i,t} \tag{2.20}$$

$$M_{i,t}^{(n)} = g(X_{i,t}^{(n)})b_{i,N}\Delta N_{i,t} \tag{2.21}$$

$$I_{i,t-1}^{(f)} = g(X_{i,t}^{(f)})a_{i,F}\Delta F_{i,t} \tag{2.22}$$

$$M_{i,t}^{(f)} = g(X_{i,t}^{(f)})b_{i,F}\Delta F_{i,t} \tag{2.23}$$

Equations (2.20)–(2.23) describe learning by doing for the investment and maintenance cost of non-fossil and fossil energy consumption respectively, where $I_{i,t}^{(n)}$ and $I_{i,t}^{(f)}$ are investment while $M_{i,t}^{(n)}$ and $M_{i,t}^{(f)}$ are for maintenance, function g(.) reflects technological change from accumulative consumption, $a_{i,F}, b_{i,F}$ is the capital intensity and maintenance intensity for fossil energy respectively, $a_{i,N}, b_{i,N}$, $a_{i,F}, b_{i,F}$ is the capital intensity and maintenance intensity for non-fossil energy respectively, $X_{i,t}^{(f)}$, $X_{i,t}^{(n)}$ is the accumulative consumption of fossil and non-fossil energy respectively. By van der Zwaan et al. (2002), we give

$$g(x) = g_0 x^{\upsilon-1} \tag{2.24}$$

$$l_r = 1 - 2^{\upsilon-1} \tag{2.25}$$

in which the value of the exponent $\upsilon - 1$ is the basis of the process of learning-by-doing and derives from Eq. (2.25) related to learning rate l_r.

2.2.3 GDP Spillovers

The mechanism of GDP spillovers in MRICES model is similar to Wang et al. (2010), but the bilateral structure is extended so that a multilateral GDP spillover can be investigated in climate change model. Figure 2.2 illustrates how the modeling of GDP spillovers is coupled with other parts of the model. Generally speaking, GDP of different countries work on each other, so that the emission of each country is impacted, resulting a global temperature raise which will make a feed back to productivity of each country. Thus a circular interaction from GDP spillovers to productivity via temperature raise is constructed. For data inaccessibility we modeled the GDP spillovers among China, the United States, Japan and Europe Union. The GDP influence from other regions can be expressed in residual.

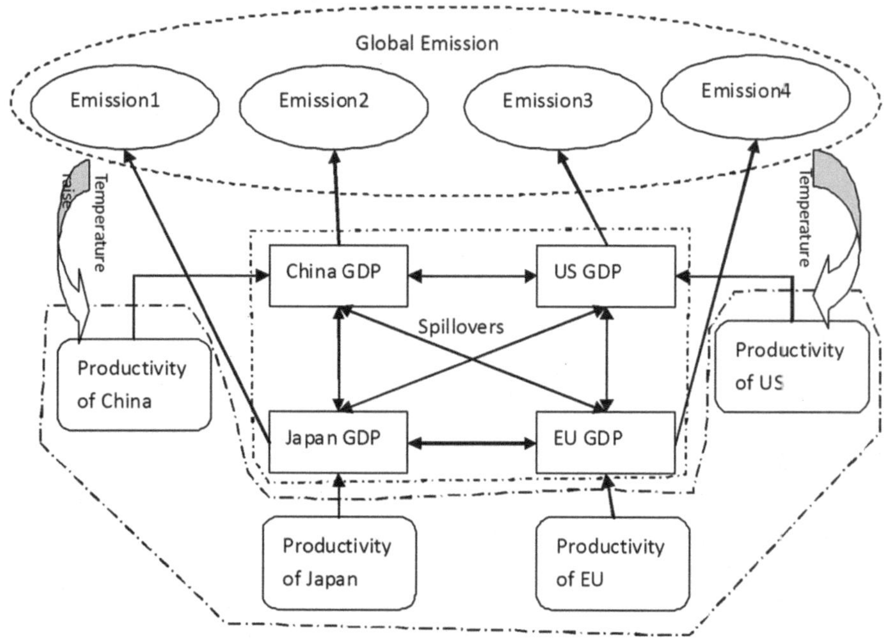

Fig. 2.2 Coupling of GDP spillover with Climate sub-system and Economic sub-system

Based on Mundell-Fleming model and Wang et al. (2010), the change of GDP is decomposed into impact from monetary policies, political policies and the GDP fluctuation of other countries, expressed in Eqs. (2.26)–(2.29).

$$\Delta q^c = v_1^c \Delta \lambda^{ca} - v_2^c \Delta (i^c - p_{+1}^c + p^c) + v_3^c \Delta q^a + v_4^c \Delta q^j + v_5^c \Delta q^e \quad (2.26)$$

$$\Delta q^a = v_1^a \Delta \lambda^{ea} - v_2^a \Delta (i^a - p_{+1}^a + p^a) + v_3^a \Delta q^c + v_4^a \Delta q^j + v_5^a \Delta q^e \quad (2.27)$$

$$\Delta q^j = v_1^j \Delta \lambda^{ja} - v_2^j \Delta (i^j - p_{+1}^j + p^j) + v_3^j \Delta q^c + v_4^j \Delta q^a + v_5^j \Delta q^e \quad (2.28)$$

$$\Delta q^e = v_1^e \Delta \lambda^{ae} - v_2^e \Delta (i^e - p_{+1}^e + p^e) + v_3^e \Delta q^c + v_4^e \Delta q^a + v_5^e \Delta q^j \quad (2.29)$$

In expression (2.26)–(2.29), superscript c, a, j, e presents variables of China, the US, Japan and EU respectively. Δq represents the deviation from the base simulation result. $\Delta \lambda$ presents the deviation from the base of real exchange. Δi presents the deviation from the base of interest rate.

Based on Eqs. (2.26)–(2.29), we can get the GDP spillover between countries (regions), shown as Eqs. (2.30)–(2.33).

$$\ln Y^c - \ln Y^c_{-1} = v^c_3(\ln Y^a - \ln Y^a_{-1}) + v^c_4(\ln Y^j - \ln Y^j_{-1}) + v^c_5(\ln Y^e - \ln Y^e_{-1}) \tag{2.30}$$

$$\ln Y^a - \ln Y^a_{-1} = v^a_3(\ln Y^c - \ln Y^c_{-1}) + v^a_4(\ln Y^j - \ln Y^j_{-1}) + v^a_5(\ln Y^e - \ln Y^e_{-1}) \tag{2.31}$$

$$\ln Y^j - \ln Y^j_{-1} = v^j_3(\ln Y^c - \ln Y^c_{-1}) + v^j_4(\ln Y^a - \ln Y^a_{-1}) + v^j_5(\ln Y^e - \ln Y^e_{-1}) \tag{2.32}$$

$$\ln Y^e - \ln Y^e_{-1} = v^e_3(\ln Y^c - \ln Y^c_{-1}) + v^e_4(\ln Y^a - \ln Y^a_{-1}) + v^e_5(\ln Y^j - \ln Y^j_{-1}) \tag{2.33}$$

In expression (2.30)–(2.33), Y^c, Y^a, Y^j and Y^e presents the GDP of China, the US, Japan and EU respectively by which the dynamic interaction between national economy is constructed with GDP spillover mechanism.

2.3 Parameter Estimation

With the model construction described above, the MRICES system was developed to simulate climate change mitigation strategies. Since some of the parameters in economic system are referenced from Eyckmans and Tulkens (2003), Nordhaus and Boyer (2000) and van der Zwaan et al. (2002), we just list the values in Appendix A.

What should be pointed out is that the descending of emission intensity from Nordhaus (1993) and Pizer (1999) is figured out on 1990s' data, however the current technological progress makes fleetly reduced emission intensity in recent years. In other words, the progress in emission intensity from Nordhaus (1993) and Pizer (1999) is underestimated for current condition. So that we estimate the current change of emission intensity based on input-output table of China and the US, which are representative for developing countries and developed countries respectively. Result indicates that the emission intensity decrease 5.3% annually in developing countries with 4.6% in developed countries. For convenience, the change of emission intensity from Nordhaus (1993) and Pizer (1999) is denoted as 'lower technological change' and the one we estimated is denoted as 'higher technological change' during the rest of the paper. We will consider each mitigation strategy under the two technological progress levels.

On the other hand, since the GDP spillover module is the new we integrate into the IAM, we determine the parameter estimation as well. Run with SPSS13.0, we get the value for GDP spillovers as Tables 2.1, 2.2, 2.3 and 2.4.

Table 2.1 Values for parameters in Eq. (2.30)[a]

Parameter	v_0^c	v_1^c	v_2^c	v_3^c	v_4^c	v_5^c	v_6^c
Value	0.004	0.067	−0.005	0.432	0.229	−1.533	1.805
Sig.	0.002	0.334	0.357	0.105	0.104	0.001	0.000

[a]Analyzed with SPSS 13.0; significance was set at the 5% level

Table 2.2 Values for parameters in Eq. (2.31)[a]

Parameter	v_0^a	v_1^a	v_2^a	v_3^a	v_4^a	v_5^a	v_6^a
Value	−0.01	−0.174	−0.271	−0.016	0.399	−0.027	0.259
Sig.	0.396	0.060	0.001	0.152	0.065	0.165	0.102

[a]Analyzed with SPSS 13.0; significance was set at the 5% level

Table 2.3 Values for the parameters in Eq. (2.32)[a]

Parameter	v_0^j	v_1^j	v_2^j	v_3^j	v_4^j	v_5^j	v_6^j
Value	0.042	−0.038	0.127	0.032	-0.082	1.011	0.154
Sig.	0.479	0.239	0.057	0.112	0.185	0.001	0.290

[a]Analyzed on SPSS 13.0; significance was set at the 5% level

Table 2.4 Values of the parameters in Eq. (2.33)[a]

Parameter	v_0^e	v_1^e	v_2^e	v_3^e	v_4^e	v_5^e	v_6^e
Value	−0.004	0.015	−0.194	0.00006	0.515	0.167	0.238
Sig.	0.102	0.436	0.000	0.746	0.000	0.053	0.014

[a]Analyzed on SPSS 13.0; significance was set at the 5% level

2.4 Assessment of Emissions Mitigation Strategies

Nowadays, a variety of mitigation strategies are being proposed (Stern 2008; BASIC 2006; Baer and Athanasiou 2008). But the quantitative influences from these strategies on national economy and global climate change are unclear, for which we can't discriminate which strategy is much better. Therefore, we assessed several mitigation strategies using our model from the perspective of effectiveness, equality and efficiency.

2.4.1 Egalitarian Allocation of Emissions Quotas

The egalitarian principle allocates emissions permits based on the population of each country. People in both developed and developing countries have equal rights to an emissions permit. However, a key problem with this principle is that industrialization occurred much earlier in developed countries than in developing countries. Tremendous carbon emissions levels have been observed since the Second Industrial Revolution due to industrialization in developed countries. As a result, North America and Europe have produced approximately 70% of the total amount of CO_2 emissions since 1850 due to energy production, whereas developing countries have accounted for less than 25% of these emissions (Stern 2006). In developing countries with extremely low historical levels of emissions, the current level of emissions is necessary to support their current level of industrialization. Thus, it would be unfair to assign emissions permits to developing and developed countries under the egalitarian principle without considering historical emissions levels.

To evaluate the control of global CO_2 concentrations to within 500 ppm by the year 2050, 1990 and 2005 were used as the starting point for historical emissions levels and the year for allocation, respectively. The mechanism for egalitarian allocation can be described by Eqs. (2.34) and (2.35).

$$P_i^{All} = pop_{i,2005} * \frac{P_w}{pop_{w,2005}} \tag{2.34}$$

$$P_i^R = P_i^{All} - P_i^H \tag{2.35}$$

Here, P_i^{All} represents the total number of emissions permits that nation i is assigned during the period from 1990 to 2050; $pop_{i,2005}$ and $pop_{w,2005}$ represent the populations of nation i and the world in 2005, respectively; P_w is the total number of emissions permits the entire world can use; P_i^R denotes the number of residual emissions permits of nation i from 2006 to 2050; and P_i^H is the level of historical emissions during the period from 1990 to 2005. The specific levels of historical emissions and residual emissions for several nations (regions) are shown in Fig. 2.3.

The business-as-usual (BAU) emissions level of each nation can be estimated based on MRICES, and the emissions gaps between BAU emissions demands and residual emissions quotas are shown in Table 2.5. Clearly, the emissions gap in the

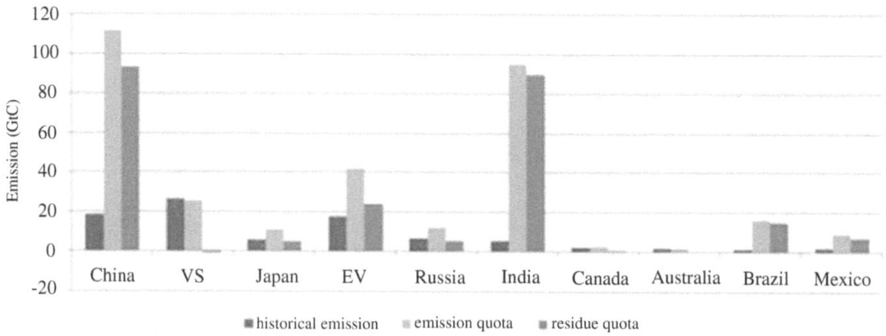

Fig. 2.3 The emissions quotas for each country (region) in the egalitarian strategy

Table 2.5 Emissions trading for each nation (region) in the egalitarianism strategy[a]

	China	US	Japan	EU	FSU	ROW
BAU emissions until 2050	95.93	81.77	16.95	53.08	44.20	164.31
Residual emissions quotas	93.39	−0.96	5.08	24.11	5.3	239
Emissions gaps	2.54	82.73	11.87	28.97	38.90	−74.68

[a]The year 1990 is used for a historical emissions starting point (units: GtC)

Table 2.6 Simulation results for the egalitarianism strategy

	Lower technological change	Higher technological change
	Loss of accumulated GDP by 2050	Loss of accumulated GDP by 2050
China	−0.013%	0.004%
US	−0.7%	−0.5%
Japan	−0.34%	−0.12%
EU	−0.33%	−0.019%
FSU	−2.8%	−1.9%
ROW	0.03%	0.016%
Temperature rise by 2050	1.48 °C	1.42 °C
Temperature rise by 2100	2.31 °C	2.02 °C
CO_2 concentration by 2050	426.7 ppm	414.6 ppm
CO_2 concentration by 2100	508.89 ppm	450.37 ppm
Change of world utility by 2050	−0.18%	−0.01%
Change of world utility by 2100	0.88%	1.55%

US is extremely large, at approximately 82.73 GtC, and the US must develop substantial emissions reduction measures or purchase emissions permits from other nations. In the case of China, the emissions demands during the period from 2008 to 2050 are almost equal to its residual emissions quotas, which imply that the emissions reduction burden for China will be relatively less than that of other developed nations if the egalitarian allocation strategy is instituted. In Table 2.5, it appears that the FSU must buy many emissions permits (Table 2.5), but this is not the case in actuality because there is 'hot air'[1] in the FSU that can alleviate the pressure of emissions reduction.

To evaluate the influence of the egalitarianism strategy on global climate change, a policy simulation based on MRICES was developed. The results are summarized in Table 2.6, and the changes in CO_2 concentrations are depicted in Fig. 2.4.

[1]'hot air': when the baseline emissions are below their entitlements, the amounts of the abundant emissions rights are referred to as 'hot air'.

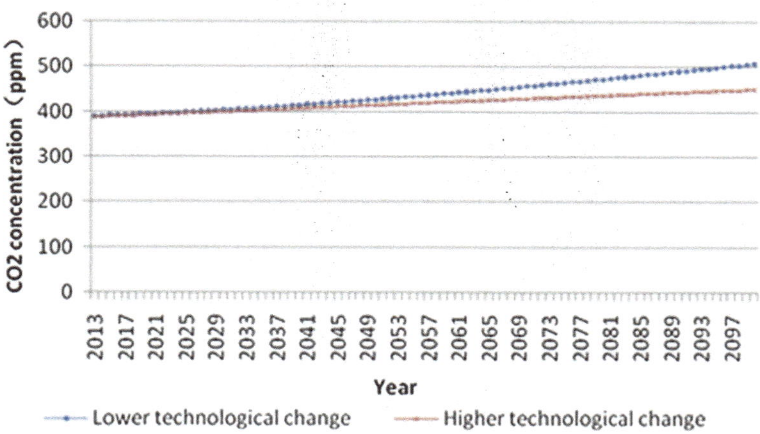

Fig. 2.4 Changes in CO_2 concentrations under the egalitarianism strategy, using 1990 as the baseline for historical emissions levels

With lower levels of technological change, the egalitarianism strategy can control the CO_2 concentration to within 500 ppm by 2050, but the temperature change by 2100 would surpass the target set by the EU. These results illustrate that a CO_2 concentration within 500 ppm is not consistent with the 2 °C target by 2100. The CO_2 concentration limit at 2050 must be further lessened, which means that countries must reduce emissions even more if this target is to be met by 2100. Notably, however, if the levels of technological change increase, both the 2050 and 2100 targets will be attained.

From the perspective of economic influence, the FSU will suffer from GDP loss more than other nations because the reduction rate in the FSU is relatively higher when the model does not consider the 'hot air', which leads to a stronger economic influence. Compared with the world accumulated utility by 2050 and 2100 under BAU scenario, it is found that the change of accumulated utility by 2050 is negative but positive by 2100 either with lower or higher technological change. So in the long-run egalitarian strategy is beneficial for utility improvement.

Thus, the egalitarianism strategy, using 1990 as the baseline for historical emissions levels, can achieve the 2050 target of controlling CO_2 concentration to within 500 ppm and the 2100 target of controlling the global temperature change to within 2 °C. Based upon the results of this strategy, developed nations must make a stronger commitment to activities that counteract the climate change caused by their tremendous historical levels of emissions. Due to the emissions gaps that developed nations face, these nations may need to purchase emissions rights instead of focusing on domestic emissions reduction measures to alleviate the average mitigation cost.

2.4.2 UNDP Strategy

In 2008, the UNDP proposed a mitigation strategy in which developed countries were required to reduce their emissions levels by 80% relative to their 1990 levels, and developing countries had to reduce their emissions levels by 20% relative to their 1990 emissions levels (UNDP 2008).

Using a MRICES policy simulation, we found that the global temperature will rise approximately 1.79 and 1.75 °C under lower and higher levels of technological change, respectively, under the UNDP strategy. This result indicates that the UNDP strategy would be effective for controlling global climate change. In particular, under higher levels of technological change, each country could attain its mitigation target with much lower reduction rates than those under lower levels of technological change.

With regard to the changes in global CO_2 concentrations, the trends under the two levels of technological change develop differently, as shown in Fig. 2.5. Specifically, under lower levels of technological change, the trend in CO_2 concentrations is U-shaped, which indicates that although the concentration would fall in the beginning of the mitigation period, it would then begin to increase after approximately 2045. Conversely, the trend in CO_2 concentrations under higher levels of technological change converges by the end of the century, which indicates that global emissions will be controlled at a certain level in the long-term. Thus, a higher level of technological change is distinctly better for accomplishing climate protection.

The accumulated GDP losses in each country resulting from reaching the targets described in the UNDP strategy are listed in Table 2.7. The GDP losses in developing countries are higher than those in developed countries, indicating that developing countries must invest more relative to their ability to pay. However, the

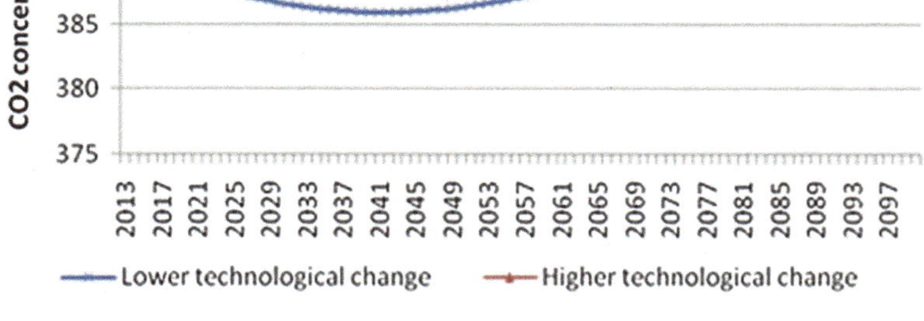

Fig. 2.5 Changes in CO_2 concentrations under the UNDP strategy

Table 2.7 Simulation results for the UNDP strategy

	Lower technological change	Higher technological change
	Loss of accumulated GDP by 2050	Loss of accumulated GDP by 2050
China	1.46%	0.52%
US	0.92%	0.46%
JAPAN	0.36%	0.17%
EU	0.38%	0.01%
FSU	1.79%	0.36%
ROW	0.96%	1.4%
Temperature rise by 2050	1.43 °C	1.48 °C
Temperature rise by 2100	1.79 °C	1.75 °C
CO_2 concentration by 2050	386.5 ppm	392.9 ppm
CO_2 concentration by 2100	402.9 ppm	393.5 ppm
Change of world utility by 2050	−0.72%	−0.44%
Change of world utility by 2100	−0.07%	0.26%

situation associated with GDP loss in developing countries becomes less severe when their emissions decline based on a higher level of technological change. Therefore, transferring advanced technologies from developed countries to developing countries can help to release the mitigation burden on developing countries.

Although the influence of emissions reduction on the economy is relatively tempered, this strategy is still far from appropriate as a basis for global action. The emissions per capita by 2050 varies among countries; the highest per capita level in the US is approximately 4.7TCO$_2$e, which is one-fold the world average, as shown in Fig. 2.6.

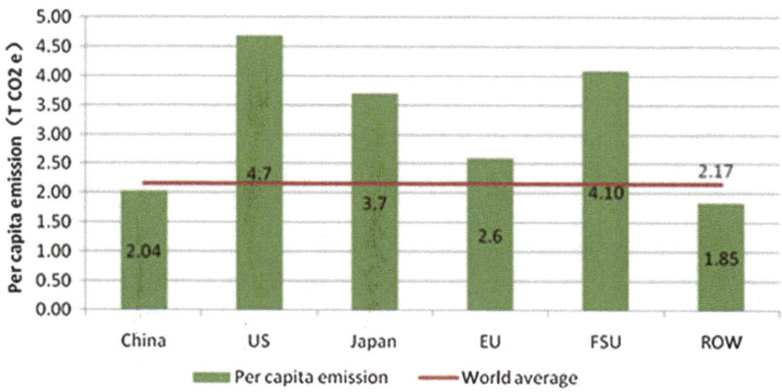

Fig. 2.6 Emissions per capita by 2050 under the UNDP strategy

Thus, the UNDP Strategy can address global climate change to a significant extent, but developing countries must absorb higher costs for mitigation and gain lower emissions per capita than developed countries, which will result in a new inequality in emissions rights and development rights throughout the world. The change of world utility by 2050 and 2100 is also undesirable when compared with that of egalitarian strategy, which means the world cost more for climate change. The key fault of the UNDP strategy is that developing countries are required to participate in emissions reduction simultaneously with developed countries. Developing countries cannot manage the implementation of emissions reduction measures and the elimination of poverty at the same time.

Based on our assessment, there should be differentiated obligations for developed and developing nations, with the objective of sustaining the equity rights of economic development. Therefore, two key issues should be emphasized when developing a global emissions mitigation strategy. Developed nations must assume more responsibility in global emissions reduction. Furthermore, developed nations must implement substantial emissions reduction measures earlier than developing nations so that developed nations will have time to achieve economic development and further alleviate the effects of climate change on the economies of developing nations.

2.4.3 Copenhagen Accord

Based on the Copenhagen Accord, Annex I parties should strengthen their emissions reduction measures, whereas non-Annex I parties should develop nationally appropriate mitigation actions. The mitigation action targets for 2020 proposed by the five countries (regions), which were modeled using MRICES, are listed in Table 2.8.

Because no strategy has been proposed for emissions reduction after 2020, we assume that each country will continue the mitigation actions listed in Table 2.8 until 2100 to explore the consequences on climate change. Based on a policy simulation using MRICES, the increase in global temperature was found to be

Table 2.8 Mitigation actions for 2020 in Copenhagen Accord

	Mitigation target	Base year	Base value	Target value
China	Reducing the emissions intensity per unit of GDP in 2020 by 40–45%	2005	7.99 t/10,000 $	4.39 t/10,000 $
US	Reducing emissions 17% by 2020	2005	1.57 GtC	1.30 GtC
EU	Reducing emissions 30% by 2020	1990	1.20 GtC	0.84 GtC
Japan	Reducing emissions 25% by 2020	1990	0.29 GtC	0.22 GtC
Russia	Reducing emissions 25% by 2020	1990	1.03 GtC	0.75 GtC

approximately 1.2 °C by 2020, 1.9 °C by 2050 and 2.8 °C by 2100. Clearly, this result exceeds the target to limit the rise in global temperature to within 2 °C by 2100. Therefore, much stronger mitigation actions must be taken after 2020; otherwise, combating climate change will become rather difficult if the temperature increases 1.9 °C by 2050.

From the perspective of emissions intensity changes, a relatively even reduction rate was revealed among the modeled countries. China exhibits a 45% reduction in emissions intensity from 2005 levels by 2020, and the US, the EU, Japan and the FSU exhibit 47%, 46%, 57% and 49% reductions in emissions intensity, respectively. However, emissions reduction is not easy for any country, and technological development is crucial for all countries; in particular, technological transfer from developed to developing countries is essential.

2.4.4 A Strategy to Achieve the 2 °C Target

On the basis of the "common but differentiated responsibilities" principle, developed countries are asked to develop more emissions reduction measures at an earlier time to address global climate change. Therefore, to achieve the 2 °C target by 2100, we propose that developed countries reduce emissions by 80% below 1990 levels by 2050, that the FSU reduce emissions by 50% below 1990 levels by 2050 and that China and the ROW reduce emissions by 28% and 20%, respectively, below 2005 levels. Policy simulations show that the strategy we proposed can maintain the temperature rise below 2 °C by the year 2100.

Using this strategy, we compared the decreases in carbon intensity of each country by 2050 and 2100, as shown in Fig. 2.7. China will exhibit an 89.9% reduction in carbon intensity by 2050 relative to 2005, which is close to that of developed countries, and the decrease in carbon intensity of the ROW will be reduced by 79.6% by 2050, which is relatively lower than those of other countries. However, the decrease in carbon intensity will be nearly equal among all countries

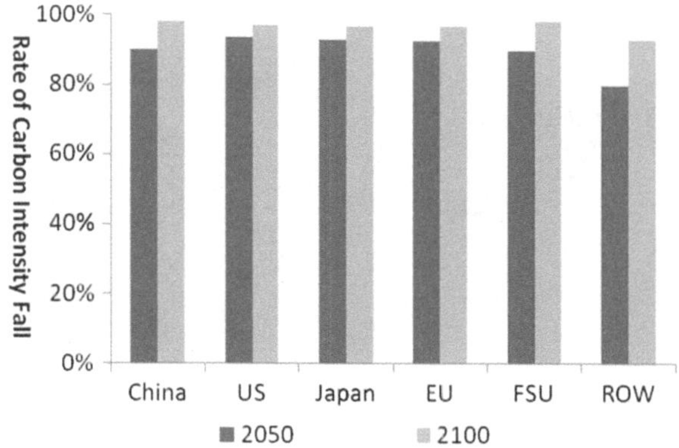

Fig. 2.7 Comparison of the declines in carbon intensity by 2050 and 2100

by 2100, and the decreases in China and the FSU will even surpass those of developed countries. Thus, based on the changes in carbon intensity, the emissions reduction efforts in developed and developing countries are equal.

2.5 Conclusions and Discussion

As we approach the post-Kyoto era, there are an increasing number of mitigation strategies to address climate change. This study sought to develop an assessment strategy for these strategies based on the objectives of effectiveness, equality and efficiency. For this purpose, a climate policy model with GDP spillovers was presented. Based on an extension of the Mundell-Fleming model, we simulated the economic interactions among China, the US, the EU and Japan with a GDP spillover mechanism. Consequently, emissions reduction measures developed in one country will not only influence that country's economy but also the economies of the other countries via GDP spillover.

Based on assessments on the egalitarian allocation strategy, the UNDP strategy and the Copenhagen Accord using MRICES, it is found that the egalitarian strategy is helpful for developed countries to get more emission permits.

On account of equality, efficiency and effectiveness, there are several principles which we have pay more attentions to when making global emission reduction strategies. Firstly, the principle of common but differentiated responsibility must be insisted, so that developed countries can contribute more in emission reduction. Based on this principle, cutting 80% relative to 1990 levels by 2050 in developed countries can be considered as the precondition for emission reduction in developing countries. Secondly, it's not practical to take the same year as the baseline for emission reduction in both developed and developing countries, otherwise the reduction target will be too challenging for developing countries. Alternatively, 1990 and 2005 can be base years for developed and developing countries respectively. Thirdly, promoting advanced technologies from developed countries to developing countries is one of the ways to combat global climate change more effectively and efficiently.

Appendix A

See Appendix Table A.1.

Table A.1 Values for parameters in Eqs. (2.1)–(2.25)

Parameter	Equations	Value						References
		China	US	Japan	EU	FSU	ROW	
α	Equation (2.1)	0.3	0.25	0.25	0.25	0.3	0.25	Eyckmans and Tulkens (2003)
D_0	Equation (2.2)	0.1371	0.0992	0.1057	0.1057	0.0771	0.144	Eyckmans and Tulkens (2003)
$b_{i,1}$	Equation (2.2)	0.15	0.07	0.05	0.05	0.15	0.10	Eyckmans and Tulkens (2003)
$\gamma_{i,a}$	Equation (2.3)	0.033	0.165	0.165	0.165	0.033	0.033	Nordhaus and Yang (1996)
$\delta_{i,a}$	Equation (2.3)	0.006	0.005	0.005	0.005	0.006	0.006	Nordhaus and Yang (1996)
$\delta_{i,k}$	Equation (2.4)	0.1	0.1	0.1	0.1	0.1	0.1	
η_i	Equation (2.5)	0.3	0.25	0.25	0.25	0.3	0.25	
ρ_i	Equation (2.7)	0.015	0.015	0.015	0.015	0.015	0.015	Eyckmans and Tulkens (2003)
τ_i	Equation (2.7)	0.198	0.198	0.198	0.198	0.198	0.198	
$\sigma_{i,t}$	Equation (2.8)	0.0007	0.0001	0.00006	0.0001	0.0007	0.0003	
$\gamma_{i,t}$	Equation (2.10)	0.033	0.165	0.165	0.165	0.033	0.033	Nordhaus and Yang (1996)
$\sigma_{i,t}$	Equation (2.12)	0.00075	0.00015	0.000068	0.00012	0.00072	0.00029	
ς_i	Equation (2.13)	0.22	0.22	0.11	0.2	0.22	0.51	
$\lambda_{i,FN}$	Equation (2.15)	0.93	0.88	0.83	0.87	0.90	0.88	
$a_{i,N}$	Equation (2.20)	3	2	2	2	3	3	van der Zwaan (2002)
$b_{i,N}$	Equation (2.21)	9	8	8	8	9	9	van der Zwaan (2002)
$a_{i,F}$	Equation (2.22)	3	3	3	3	3	3	van der Zwaan (2002)
$b_{i,F}$	Equation (2.23)	1	1	1	1	1	1	van der Zwaan (2002)
g_0[a]	Equation (2.24)	1	1	1	1	1	1	van der Zwaan (2002)
g_0[b]	Equation (2.24)	2	2	2	2	2	2	van der Zwaan (2002)
l_r[a]	Equation (2.25)	0.07	0.07	0.07	0.07	0.07	0.07	van der Zwaan (2002)
l_r[b]	Equation (2.25)	0.1	0.1	0.1	0.1	0.1	0.1	van der Zwaan (2002)

[a]Value for learning-by-doing in fossil-fuel technologies
[b]Value for learning-by-doing in non fossil-fuel technologies

References

Baer P, Athanasiou T (2008) The right to development in a climate constrained world: greenhouse development rights framework. http://www.ecoequity.org/GDRs/GDRs_ExecSummary.html. Accessed 10 Nov 2010

BASIC (2006) The Sao Paulo proposal for an agreement on future international climate policy. http://www.basic-project.net/. Accessed 15 Sept 2010

Buonanno P, Carraro C, Galeotti M (2003) Endogenous induced technical change and the costs of Kyoto. Resour Energy Econ 25:11–34

Burniaux J, Nicoletti G, Oliviera-Martins J (1992) GREEN: a global model for quantifying the cost of policies to curb CO_2 emissions. OECD, Paris

Bussolo M, de Hoyas R, Medvedev D et al (2008) Global climate change and its distributional impacts. Paper presented to eleventh annual conference on global economic analysis, Helsinki, June

Chen W, Gao P, He J (2004) Impact of carbon mitigation on China's energy system using China MARKAL-MACRO model. J Tsinghua Univ (Sci Technol) 44(3):342–346 (in Chinese)

Dixon PB, Parmenter BR, Sutton J et al (1982) ORANI: a multisectoral model of the Australian economy. North-Holland, Amsterdam

Eyckmans J, Tulkens H (2003) Simulating coalitionally stable burden sharing agreements for the climate change problem. Resour Energy Econ 25:299–327

Golombek R, Hoel M (2003) Climate policy under technology spillover. Environ Resour Econ 31 (2):201–227

Grubb MJ, Hope C, Fouquet R (2002) Climatic implications of the Kyoto Protocol: the contribution of international spillover. Clim Change 54:11–28

Jha R (2009) The global financial crisis and short-run prospects for India. ASARC working paper

Jiang K, Hu X (2002) Countermeasure analysis on climate change. Beijing, Working paper of Energy Research Institute National Development and Reform Commission (in Chinese)

Kelly DL, Kolstad CD (1999) Integrated assessment models for climate change control. In: Folmer H, Tietenberg T (eds) The international yearbook of environmental and resource economics 1999/2000, Edward Elgar, Cheltenham, pp 171–197 (1999)

Kemfert G (2005) Induced technological change in a multi-regional, multi-sectoral integrated assessment model. Ecol Econ 54:293–305

Link PM, Tol RSJ (2004) Possible economic impacts of a shutdown of the thermohaline circulation: an application of FUND. Portuguese Econ J 3(2):99–114

Ligang Liu (2009) Impact of the global financial crisis on China: empirical evidence and policy implications. China World Econ 17(6):1–23

Loulou R, Labriet M (2007) ETSAP-TIAM: the TIMES integrated assessment model. Part I: Model structure. Comput Manage Sci 5:7–40

Loulou R (2008) ETSAP-TIAM: the TIMES integrated assessment model. Part II: Mathematical formulation. Comput Manage Sci 5:41–66

Luh Y, Shih K (2006) International spillovers and east Asian growth: the experience of Japan, Korea and Taiwan. Appl Econ Lett 13(11):745–750

Manne AS, Mendelsohn R, Richels R (1995) MERGE—a model for evaluating regional and global effects of GHG reduction policies. Energy Policy 23(1):17–34

Manne AS, Richels R (2004) MERGE: a model for evaluating the regional and global effects of GHG reduction policies. http://www.stanford.edu/group/MERGE. Accessed 3 Nov 2010

McKibbin WJ, Wilcoxen PJ (1993) The global consequences of regional environmental policies: an integrated macroeconomic, multi-sectoral approach. In: Kaya Y, Nakicenovic N, Nordhaus WD et al (eds) Costs, impacts and benefits of CO_2 mitigation, IIASA, Austria

Nijkamp P, Wang S, Kremers H (2005) Modeling the impacts of international climate change policies in a CGE context: the use of the GTAP-E model. Econ Modell 22:955–974

Nordhaus WD (1990) A general equilibrium model of policies to slow global warming. In: Wood D (eds) Economic models of energy and environment, Proceedings of a workshop, Washington, D.C

Nordhaus WD (1991) The slow or not sloe: the economics of the Greenhouse Effect. Economic J 101:920–937

Nordhaus WD (1993) Rolling the 'DICE': an optimal transition path for controlling greenhouse gases. Resour Energy Econ 15:27–50

Nordhaus WD, Boyer J (2000) Warming the world: economic models of global warming. MIT Press, Cambridge, Mass

Nordhaus WD, Yang Z (1996) A regional dynamic general-equilibrium model of alternative climate-change strategies. Am Econ Rev 86(4):741–765

OECD (1992) Global effects of the European carbon tax. Economics Department Working Papers No. 125

Pizer WA (1999) The optimal choice of climate change policy in the presence of uncertainty. Resour Energy Econ 21:255–287

Rao S, Keppo I, Riahi K (2006) Importance of technological change and spillover in long-term climate policy. Energy J 27:123–139

Stern N (2006) Stern review on the economics of climate change. Report to the Prime Minister and the Chancellor of the Exchequer on the economics of climate change, London, United Kingdom

Stern N (2008) China in the world. Speech in Tsinghua, 23 Oct

UNDP (2008) Fighting climate change: human solidarity in a divided world. Human development report 2007/2008. http://hdr.undp.org/en/reports/global/hdr2007-2008/. Accessed 15 Sept 2010

van der Zwaan BCC, Gerlagh R, Klaassen G et al (2002) Endogenous technological change in climate change modeling. Energy Econ 24:2–19

Wang Z, Hu Q, Zheng Y (2002) Simulating the impact of the climate protection expenditure on China's economic security. ACTA Ecologica Sinica 22(12):2238–2245 (in Chinese)

Wang Z, Li HQ, Wu J, Gong Y et al (2010) Policy modeling on the GDP spillovers of carbon abatement policies between China and the United States. Econ Modell 27:40–45

Wang Z, Zheng Y, Jiang Y (2004) Dynamic macroeconomic modeling and analysis of CO_2 abatement. ACTA Ecologica Sinica 24(7):1508–1513 (in Chinese)

Weeks J (2009) The impact of the global financial crisis on the economy of Sierra Leone. Int Policy Center Incl Growth 9:1–32

Chapter 3
The Impact of Sea Level Rise

3.1 Introduction

Under the influence of global climate change, countries around the world are actively involved in addressing climate change; While worldwide uniform reduction program has not yet formed, but scholars have made various proposals, such as Nordhaus (2007), Stern (2006), Sørensen (2008), UNDP (2008), Wang et al. (2012a, b). For these policy programs tend to focus on the feasibility study to assess the economic losses resulting emissions while ignoring various emissions reduction programs need to been assess.

Back in the 1990s, Hoozemans et al. (1993) Under the global sea level rise vulnerability analysis population, coastal wetlands and food production on a global scale development of the appropriate methods; Darwin et al. (2001) estimated the impact of sea level rise on economic perspective; Nicholls et al. (2008) use FUND model to estimate the state of the West Antarctic ice sheet melting after the dramatic impact on global sea-level rise resulting in a period of abrupt sea level rise provide a reference value, but did not discuss the different emission reduction policy on the impact of the loss of sea-level rise. Based on the latest research emphasizes the IAM assessment of sea level rise (Joshi et al. 2013), but still not linked to emissions reduction policy.

According to this trend, this chapter attempts to estimate sea level change losses, them result from various emission reduction policies and programs. Its technology roadmap is to simulate the effect of the countries (regions) under sea-level rise in the future, and make a different impact on the economic assessment of sea level change. The paper chose the famous Stern program, Nordhaus programs and Wang et al. (2012a, b) steady growth policies made a comparative analysis.

Z. Wang et al., *Integrated Assessment Models of Climate Change Economics*, DOI 10.1007/978-981-10-3945-4_3

3.2 Model and Data

In this chapter, we used an updated version of MRICES (Multi-factors dynamic Regional Integrated model of Climate and Economy System) model developed by Wang et al. (2012a, b), which is an improved version of RICE. Compared with RICE, MRICES increased the international economic interaction; the model structure can be divided into macroeconomic performance, climate system response, endogenous technological progress and multinational GDP spillover module. MRICES model will be divided into four categories countries worldwide, which are high-income countries, middle income countries; lower-middle-income countries, low-income countries, and taking into account China, USA, Japan, and the EU. Greater influence national, independent modeling, it will be subdivided into eight countries or regions.

EMRICES (Enhanced Multi-factors dynamic Regional Integrated model of Climate and Economy System) is an update version of MRICES, and it can analyze the pros and cons of various global cooperation abatement programs, and suggested mitigation policies effective program called global growth, the sea-level changes, and add Russia and India in the global regions, thus we divide the world into ten regions. In particular, CGE model is adopted to simulate China's economic growth in EMRICES or other region. This chapter mainly attempts to use EMRICES to calculate the effect of the rise in sea level caused by global warming, and used to assess the impact on the respective region due to a carbon tax, environmental tax on the economy.

In fact, EMRICES model was added the level rise equation (Vermeer and Rahmstorf 2009) to MRICES (Wang et al. 2012a, b). The main equation includes:

Atmospheric carbon cycle functions:

$$dM/dt = E_C - dV/dt - dSo/dt - dO/dt \tag{3.1}$$

Terrestrial carbon cycle function:

$$dV/dt = Npp_t - \iota_t V_t \tag{3.2}$$

$$dSo/dt = \varepsilon_s \iota_t V_t - \delta_s So_t \tag{3.3}$$

Ocean carbon cycle function:

$$dO/dt = \sigma_o((M_t - M_{t_0}) - \upsilon_o(O_t - O_{t_0})) \tag{3.4}$$

Temperature change function:

$$dT/dt = \chi \ln(M_t/M_0) - vT_t \tag{3.5}$$

Where M, V, Npp, So, O, E_C, T represent Atmospheric carbon content, carbon content of vegetation, vegetation carbon sequestration, soil carbon, the carbon content of marine human activity carbon emissions, the global average temperature,

In order to assess the various the sea level rise situation under emission reduction policies and programs, the paper added modules in the original sea-level rise climate model equations (Eq. 3.6), describing the global sea level rise, so will sea level rise and global warming organically linked:

$$dH/dt = a(T_t - T_0) + bdT/dt \tag{3.6}$$

Where H is the sea level, T is the global average temperature,T_0 is global average temperature initial value, a and b are constants.

Further, we refer FUND (Climate Framework for Uncertainty, Negotiation and Distribution) model. The current FUND model the impact of the rise in sea level is divided into four parts: 1. land loss; 2. wetland loss; 3. migration costs; 4. The protection costs. It was shown in Fig. 3.1 with detailed flow.

First, the area of land loss due to sea level rise is expressed as:

$$D_{t,r} = \min[\delta_r(S_t^{\gamma_r} - S_{t-1}^{\gamma_r}), \vartheta_r] \tag{3.7}$$

Where $D_{t,r}$ is the area r of land loss in the area of time t, δ_r is a function of rising sea levels, which means that the loss of sea level rise per unit of land (square kilometers/meters) lead, γ_r is a model based on elevation data correction parameter,

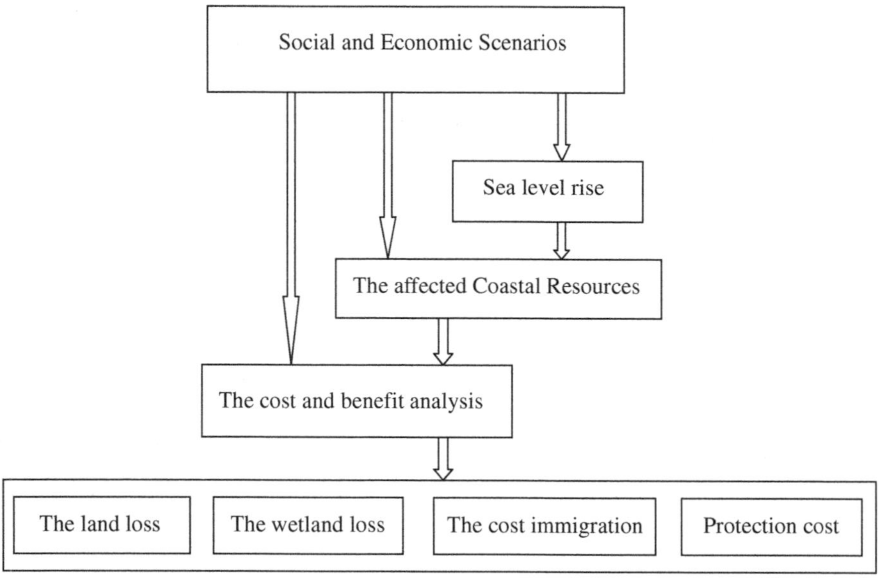

Fig. 3.1 The flow of calculating sea level effect of FUND model

S_t represents the elevation at time t of sea level, ϑ_r is the largest amount of land loss caused by sea-level rise, r is equal to the area of the area in 2000.

Land price is expressed as:

$$VD_{t,r} = \varphi(\frac{Y_{t,r}/A_{t,r}}{YA_0}) \tag{3.8}$$

Where $VD_{t,r}$ means that the price of land per unit area, $Y_{t,r}$ means that GDP, $A_{t,r}$ represents the area, φ is the argument here is equal to four million U.S. dollars/sq km, $YA_0 = 0.635$ (U.S. \$ million/km^2). Loss of land value is equal to the loss of land area multiplied by the price of land.

Loss of wetland area of sea-level rise is calculated as follows:

$$W_{t,r} = \omega_r^s \Delta S_t + \omega_r^M P_{t,r} \Delta S_t \tag{3.9}$$

Where $W_{t,r}$ is the loss of wetland area, ω_r^s and ω_r^M denote the unit due to sea level rise and coastal wetland loss caused by compaction parameters.

Wetlands price is expressed as

$$VW_{t,r} = \alpha(\frac{y_{t,r}}{y_0})^\beta (\frac{d_{t,r}}{d_0})^\gamma (\frac{W_{1990,r} - W_{t,r}^C}{W_{1990,r}})^\delta \tag{3.10}$$

Where $VW_{t,r}$ means that the price per unit area of wetlands, $y_{t,r}$ is the per capita income for the region r at time t, $d_{t,r}$ is the population density, $W_{1990,r}$ is the initial value of the wetland area, here in the region r wetland area in 1990, $W_{t,r}^C$ is the value of the cumulative loss of wetlands. α is a constant, β, γ are elastic coefficients. Loss of wetland area multiplied by the price would get the economic value of wetlands losses caused by the loss. Again, the level of protection of human response to the loss of sea-level rise caused by expanded expressed as:

$$L_{t,r} = \max[0, 1 - \frac{1}{2}(\frac{PC_{t,r} + WL_{t,r}}{DL_{t,r}})] \tag{3.11}$$

Where $L_{t,r}$ is the protection level, which means that by accounting for the protection of the coastline, to determine based on cost-benefit analysis. $PC_{t,r}$ is the net present value of the cost of coastal protection, $WL_{t,r}$ is a result of the net present value of wetland loss with coastal protection, $DL_{t,r}$ means that the net present value of land loss without protective measures.

$$PC_{t,r} = \sum\nolimits_{t=1}^{\infty} (\frac{1}{1+\rho+g_{t,r}})^t PC_0 = \frac{1+\rho+g_{t,r}}{\rho+g_{t,r}} PC_0 \tag{3.12}$$

$$WL_{t,r} = \sum\nolimits_{t=1}^{\infty} t(\frac{1}{1+\rho+g_{t,r}})^t WL_0 = \frac{1+\rho+g_{t,r}}{(\rho+g_{t,r})^2} WL_0 \tag{3.13}$$

$$DL_{t,r} = \sum_{t=1}^{\infty} t(\frac{1+g_{t,r}}{1+\rho+g_{t,r}})^t DL_0 = \frac{1+\rho+g_{t,r}}{\rho^2} DL_0 \qquad (3.14)$$

Where PC_0, WL_0, DL_0 is the average annual cost of protection, wetland loss and land value of the initial value of the loss, here considered as a constant, respectively; ρ and $g_{t,r}$ denote rate of time preference and revenue growth, where reference FUND model, ρ is valuing as 1%. Finally, the number of population migration due to the loss of land due to population density is expressed as the area multiplied by the loss of land value, the unit cost of migration is expressed as the area three times the per capita costs.

We started in 2009 as the base year to calculate the impact of future climate change and to bring. The paper divided the world to ten regions. The main units of the following variables: GDP, capital stock and capital investment, land/wetland loss values, unit: one billion U.S. dollars (B$ 2000 constant prices); Population, Unit: million (M); greenhouse gas emissions, atmospheric carbon content, unit: one billion tons of carbon (GtC); atmospheric CO2 concentration, unit: parts per million (ppm); surface temperature changes, unit: Celsius (°C); sea-level rise, unit: centimeters (cm); land/loss of wetlands, unit: square kilometer (km^2).

3.3 A Group Reduce Emissions Scheme Setting

In this chapter,we discuss a variety of policy programs, namely BAU program, Stern program, as well as the WZW program, it is a steady growth program of the global program (Wang, Zhang, Wu), BAU program is as a benchmark comparison variety programs.

Stern program is more radical reduction programs; it is that we should take immediate and effective action to reduce carbon dioxide and other greenhouse gas emissions. Stern is set in accordance with the program by 2050 to cut global carbon emissions by 50% to 1990, and thereafter, national emissions to 2100 levels by 2050 than emissions decreased by 50%.

As the Stern program set more stringent emission reduction targets, China and other developing countries is not enough to be fair and reasonable. Nordhaus proposed progressive approach to reduce emissions in the next period of time, rather than an extremely fast way. For Nordhaus program, this chapter is set: 2050, high-income countries, according to Stern reduction programs, upper middle income countries to reduce by 30% relative to 1990, China, India, 20% reduction in emissions by 2005 based on, middle and lower income countries to maintain emission levels by 2020, low-income countries without emission reduction; after addition of low-income countries do not have to keep 2050 emissions reduction beyond the rate of reduction.

World total emissions from 2020 to 2050, with 1990 as the baseline emissions, 80% reduction in carbon dioxide emissions by high-income countries, given Japan no longer suitable for large-scale expansion of nuclear power scale, their emission reduction targets 70%, upper-middle-income countries decreased by 50%, Russia 30%; compared to 2005 to reduce carbon emissions in China and India, 30%, lower-middle-income countries to keep carbon emissions by 2020, low-income countries do not participation in total emissions, making more opportunities for developing countries, countries in economic growth in order to maintain a balance, then to 2100, participation in the countries to maintain 2050 emissions reductions. We call this as *Effective Growth* scenario.

3.4 Result

Table 3.1 shows the carbon emissions under the BAU scenario. In the case of countries do not reduce emissions, the future carbon emissions will continue to rise, the total global carbon emissions rose from 2010 to 2100 of 8.44 GtC of 20.08 GtC.

In the BAU scenario, in 2009 the United States the largest per capita carbon emissions for each 4.7 tC China's lack of its 1/3, equivalent to only about 45% of high-income countries, but also lower than the European Union, Japan, Russia, the high-income countries. To 2100, national per capita carbon emissions are rising, the U.S. per capita carbon emissions rose to 6.9 tC, still ranked first, while China's per capita emissions have not reached the level of 2009 in the United States. Meanwhile, India, lower-middle-income countries, middle income countries and low-income countries, per capita emissions are still far below the level of high-income countries. Although each country's per capita emissions level of future large difference, but in the global climate change externality effects, but all countries bear the risk of rising sea levels common in the BAU scenario, which requires countries to have a positive commitment to the common emission reduction responsibility.

Table 3.1 BAU scenario under regional carbon emissions (Unit: GtC)

Year	China	USA	EU	Japan	Russia	India	High income countries	Medium high income countries	Medium low income countries	Low income countries
2010	2.17	1.47	1.0	0.3	0.44	0.56	0.79	1.13	0.53	0.05
2020	2.89	1.71	1.13	0.33	0.55	0.75	0.88	1.35	0.65	0.07
2030	3.51	1.92	1.26	0.37	0.68	0.96	0.98	1.56	0.79	0.08
2040	4.08	2.08	1.38	0.4	0.83	1.17	1.07	1.73	0.93	0.1
2050	4.58	2.21	1.48	0.42	0.98	1.37	1.15	1.86	1.07	0.12
2060	5.01	2.32	1.57	0.45	1.14	1.55	1.21	1.96	1.2	0.15
2070	5.37	2.41	1.64	0.47	1.3	1.71	1.26	2.04	1.32	0.17
2080	5.67	2.47	1.71	0.49	1.44	1.84	1.31	2.09	1.42	0.19
2090	5.91	2.52	1.76	0.5	1.58	1.96	1.34	2.13	1.52	0.22
2100	6.1	2.56	1.8	0.51	1.7	2.05	1.37	2.15	1.59	0.25

3.4.1 The Temperature

According to the case simulated carbon emissions in the future, you can calculate the future global average temperature rise. Figure 3.2, the simulated situation of each abatement program global average temperature rise. To2100, BAU scenario warming will reach 3.38 °C. Stern program because of its stringent emission reduction targets by 2100 global warming was 1.79 °C. In contrast, Nordhaus, program reduces the emission reduction efforts and finally to the 2100 global temperature rise of 2.05 °C; it very closer to the target 2 °C. Finally, the effective growth scenario considering the merits of the two scenarios, reaching 2 °C under the premise of the target, taking into account each country emissions targets, making the economic development of China and other middle-income countries have more space to control temperature rise in 2100 in the less than 2 °C, is 1.97 °C.

3.4.2 The Sea Level

Under the influence of temperature rise, according to formula (3.6), sea level rise is a linear function of temperature change with the temperature change rate of the interference term. Since the model is a semi-empirical model, *a* and *b* are two parameters estimated from historical data, select the data estimation results will have a greater impact, and this is the complexity of the climate performance. Simulating various scenarios of future global sea level rise is shown in Fig. 3.3. Sea level rise and temperature rise is proportional to the height of the relationship between the averages annual growth showed a slight increasing trend. BAU fastest rising program which, in 2009 compared to 2100 increased by 94.4 cm, Stern program rose 74.7 cm, Nordhaus' program rose 76.8 cm, effective growth scenario risen 76.4 cm. Expected compared to the IPCC Fourth Assessment Report, the height of sea level rise in this century will be between 18 –59 cm between papers

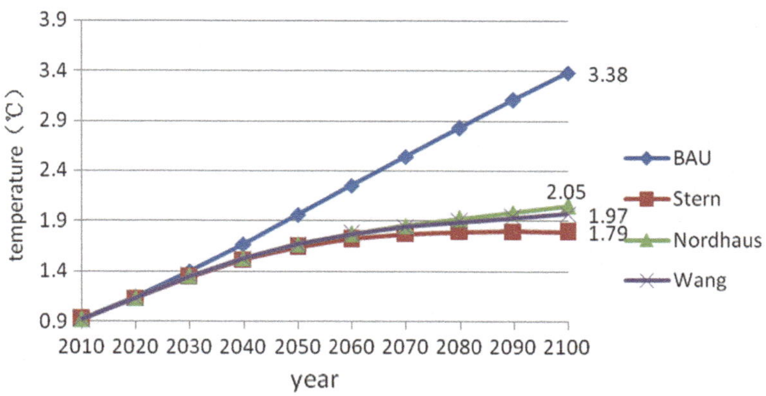

Fig. 3.2 The temperature change under scenarios

simulated the height of sea level rise over the IPCC's estimates, in 2100 an increase of nearly 1 m in height. In addition, IPCC report nor the rapid changes in the Greenland and Antarctic ice may occur into account. Semi-empirical model used in this chapter established based on historical data and estimate the parameters of the model to simulate sea level rise and global temperature rise of relations in the future. Controversial model is exactly the above simple relationship can last long, it is can't be denied that there is still a big variable, especially in the less clear the area of the Greenland ice sheet and the changes that may occur in the Antarctic.

Can also be seen from Fig. 3.3, the various options for reduction under the sea level rises were decreased, and with the global rise in temperature is proportional to the difference with the BAU Stern scheme reduces the maximum height of nearly 20 cm, the program except BAU among several other programs outside of the difference is not great.

3.4.3 The Economic Loss of Sea Level Rise

3.4.3.1 The Land Loss Without Protection

Sea-level rise is mainly led to the loss of coastal land. In Fig. 3.4 we have calculated and given the cumulative loss of each country (regional) in 2100 under resulting from sea level rise BAU scenario. With the rise in sea level, countries (regions) loss area continues to rise, especially in middle-income and low-income countries. By the 2100, The sea level will rise 94.4 cm, global land losses will up to 508,158.3 sq km, the loss of 155,636.4 sq km of wetlands. Lower middle-income countries in 2100 reached a cumulative loss of 148,104.2 sq km and 41,817.4 sq km of wetlands, is one of the largest areas of. The land loss of the smallest countries is Japan, until 2100 for the loss of 2,365.3 sq km. China land loss of primacy in

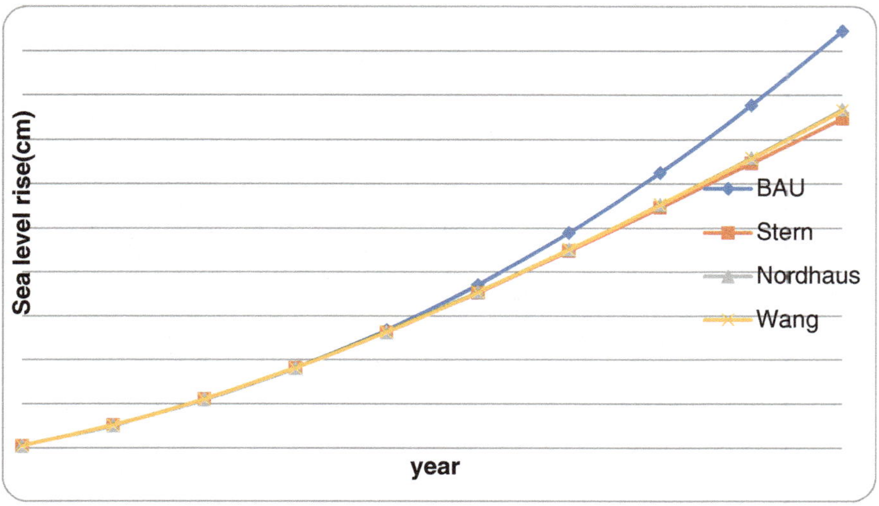

Fig. 3.3 The sea level rise under scenarios (compared to 2009)

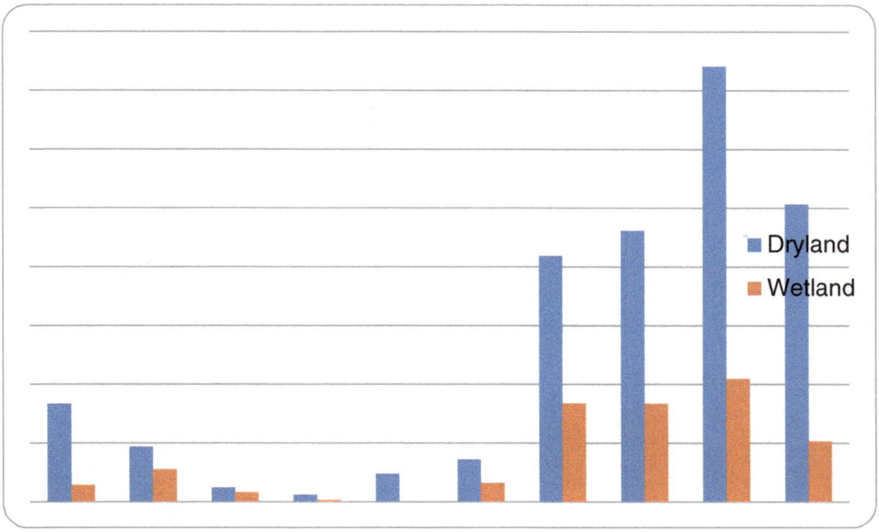

Fig. 3.4 BAU scenarios in 2100 the cumulative loss of land and wetlands contrast (sq km)

various countries around the world, and in 2100 reached 33,327.9 sq km losses; in addition, China's wetland loss less than the United States and India, and the loss of land of the United States immediately after China. The wetland loss is large for lower-income countries, followed by upper middle income countries. In short, the sea-level rise under global warming, impact on low-income countries is the most. For China, because China has a vast coastal belt is longer, the impact on China's sea level rise significantly.

Due to the loss of wetland and loss of general land has some similarity, the analysis will are to land behind a case study. Figure 3.5 shows our calculated under four scenarios of land loss comparison chart, we can see the effective growth scenario compared to the BAU scenario, Nordhaus and Stern under national programs (area) of land loss has been reduced, Stern programs and effective than under the growth scenario is slightly smaller loss, land loss and rising sea levels and how much basic proportional to the height. This shows that the case be slowing emissions rise in global temperatures, thereby reducing the rise in sea level, the area of natural land loss is also reduced. The gap between the outside of the three scenarios BAU small, it is because the difference between the program under Nordhaus, Stern programs and effective growth scenarios of sea level rise are small, so the loss of land under several scenarios difference is not great.

3.4.3.2 The Land Loss Under Protection

In response to rising sea levels, some countries can take certain protective measures. In Fig. 3.6, we have compared calculation results of the cumulative loss of land that

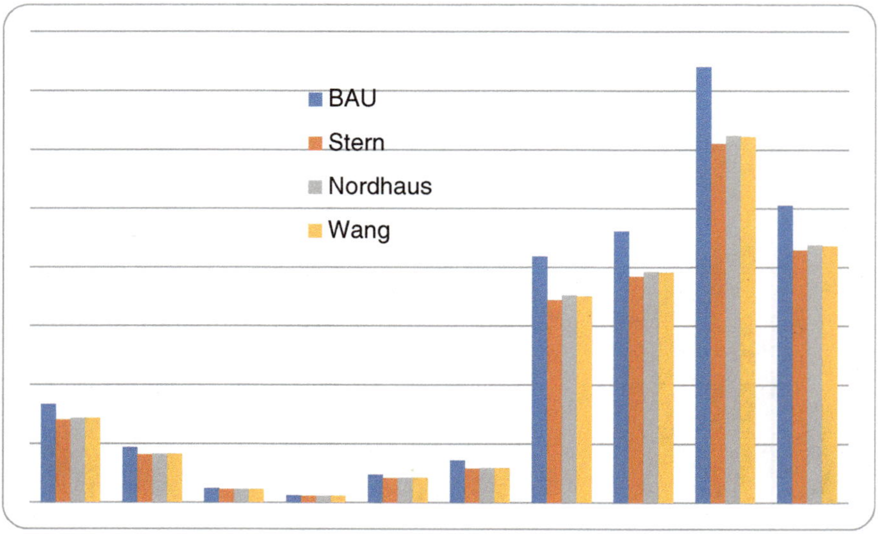

Fig. 3.5 The accumulated land loss in 2100 under scenarios

there is protection measures with there is no protection by 2100 under the BAU scenario, it can be seen from the figure, there is protection and unprotected contrast vary widely, such as China and the United States, with protective measures or without, the cumulative loss of land were 2680.7 and 3553.7 sq km. The Loss of land is larger on middle-income countries, after adding substantial protection measures, the land have also been reduced,. The effect of Japan's most significant, the protection will reduce loss compare with there is no protection to 5.6%. The protection effect of Russia is 0, which because of its large land area, wetlands and protection costs greatly exceed the value of the land value, resulting in a low level of protection of a great relationship. In the case of protective measures, four scenarios in 2100 comparing the cumulative loss of land valuesshown in Fig. 3.7.

Figure 3.7 shows, the biggest loss is the BAU scenario, the difference of other three scenarios is relatively small. In effective growth scenario, the largest loss in the lower-income countries dropped from 37, 135.4 to 29, 777.4 compare with BAU, followed by high-income countries have also fallen more than 20%, Japan, due to the loss of too little has been difficult to show in Fig. 3.7 out. The cumulative loss of Nordhaus scenario or Stern scenario is greater than the effective growth scenario.

3.4.3.3 The Economic Loss

Based on EMRICES, we calculated the loss of land and wetlands caused by sea-level rise, and estimated the economic value of direct losses and cost of the migration and protection for the population under impact of sea level rise.

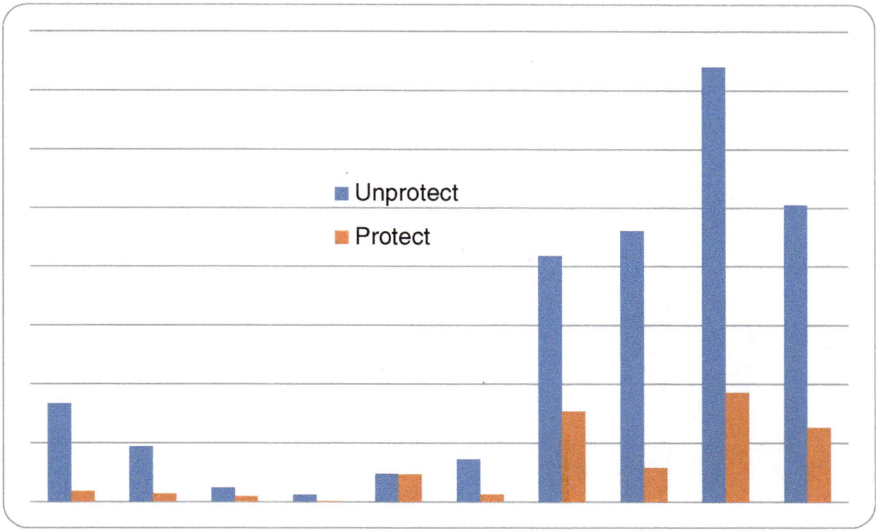

Fig. 3.6 Comparison whether the protection of land loss in 2100 under BAU scenario

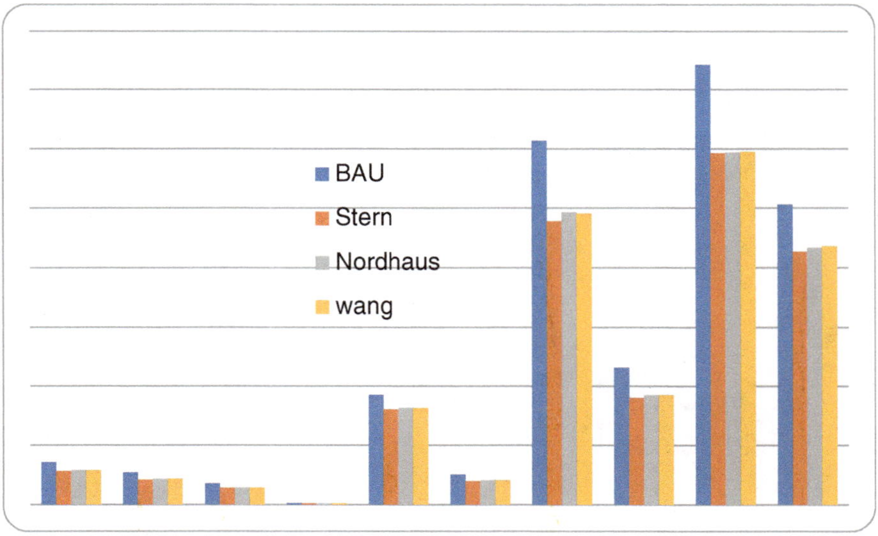

Fig. 3.7 The accumulated land loss in 2100 of different scenarios under protection

Figure 3.8 shows the four emission scenarios to 2100 countries (regions) the cumulative loss of economic value.

Under the effective growth scenario, the economic loss is also the largest lower-middle-income countries, to $341.47 billion, compared to the baseline scenario to reduce the loss of nearly 25%, the United States, the European Union and

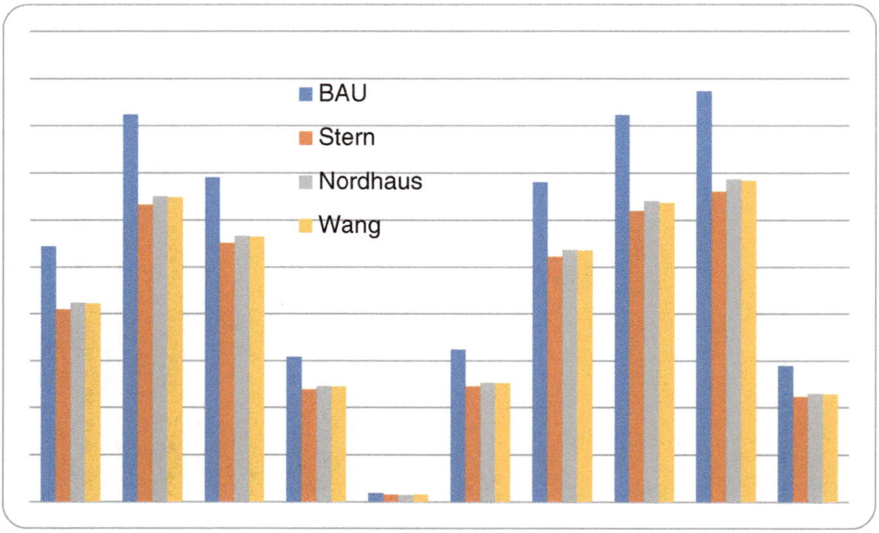

Fig. 3.8 The accumulated economic loss in 2100 under scenarios (unit: billion dollars)

other countries (regions) also reduce economic losses more than 20%. Nordhaus scenarios and Stern scenario economic losses or less, the trend remained is similar. But the losses here only consider the direct economic losses caused by sea-level rise.

Figure 3.9 shows the comparison on the economic loss under two cases under the BAU scenario. As can be seen from the Fig. 3.9, the biggest difference is that

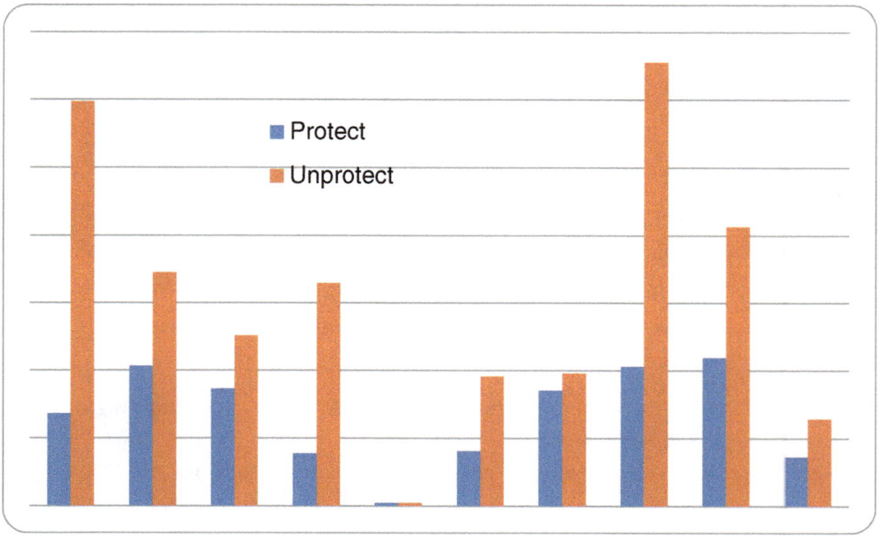

Fig. 3.9 A comparative economic value of the cumulative loss by 2100 (Unit: U.S. $ billions)

the loss of China, four times more away, followed upper middle income countries, there are also three times as many gaps; because two areas have in common is a large population, wide area, without protection and protection are two very different scenarios on population migration, the population needs to move out and move into the economic costs. Differences in population and land loss disparity gap caused economic loss are self-evident.

3.5 The Flood Area in China

Due to EMRICES is simulation systems be combined with GIS, Based on GIS, we simulated change of sea level rise submerged area in China, and Inundation of coastal areas is obtained. The study used data is SRTM DEM data. Figure 3.10

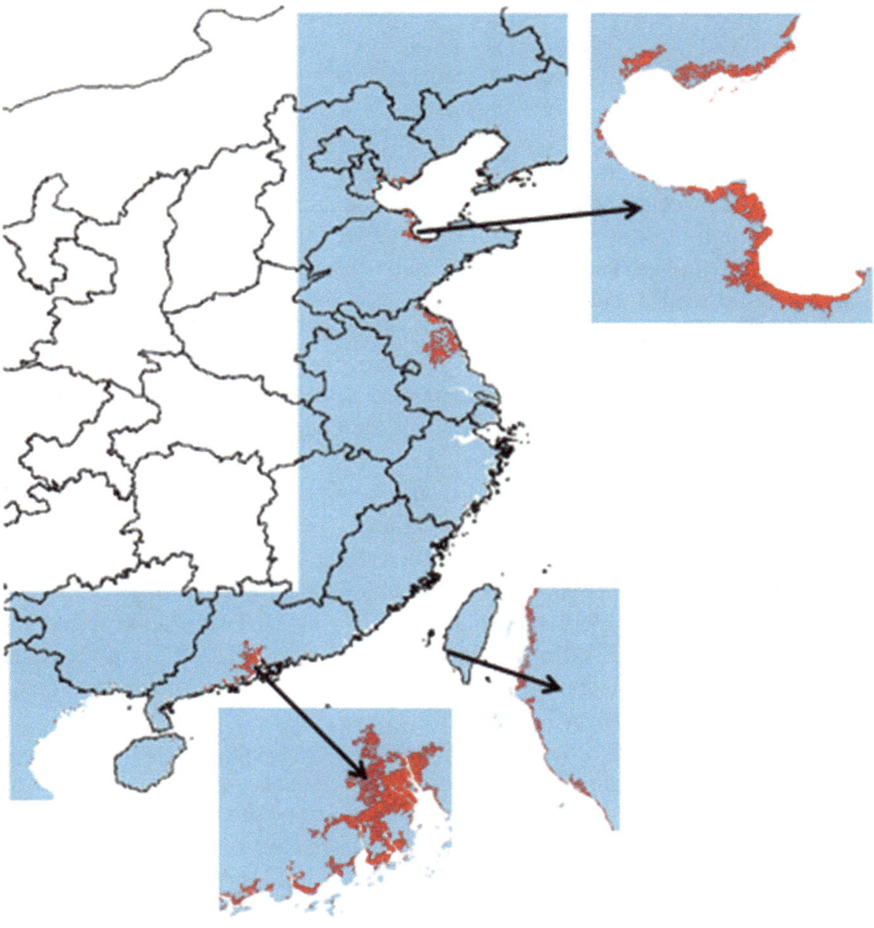

Fig. 3.10 Chinese mainland coast submerged area diagram (excluding small islands)

shows the entire Chinese coastal flooding case under BAU. (Using EMRICES simulated, we found that BAU will bring the fastest rising, the sea-level will increased 94.4 cm compared to 2009 by 2100. Therefore, we could use 100 cm to estimate the Chinese coastal flooding situation under BAU sea-level rise case).

In Fig. 3.10, the arrow points to an enlarged portion can be seen in Figure. The results discover that affected coastal part is still not small. The country is more obvious Bohai Sea, Lixiahe watershed, Shandong coastal, Tianjin and Qinhuangdao in some areas along the coast have a larger area to be flooded. Jiangsu submerged area have very reasonable place in the figure, due SRTM DEM data in some places there are holes.

With GIS spatial analysis based GIS, it were last flooded grid drawn without considering China Sea islands are submerged under the circumstances, in the case of sea-level rise of 1 m flooded area of about 28,000 sq km.

3.6 Discussion

By the simulation results of several scenarios, we found that it is impossible to achieve emission reductions in Copenhagen 2 °C consensus under the baseline scenario. In order to effectively control the rise in global temperature, corresponding reduction targets and policies must be set. Stern emission control policy programs, though could control the rise in temperature, but this scenario is more favorable to high-income countries, but it is unfair to China and other developing countries; Nordhaus scenario can control temperature lower than 2°, and appropriate efforts to reduce national emissions than high-income, has its positive side, but the carbon emissions of view there is debatable place on fair. Under the effective growth scenario can effectively control the rise in global temperatures, reducing the rise in sea level, while the pressure is relatively small, it could ensure smooth economic growth in world, reducing the countries (regions) in economic losses.

In addition, outside of the model we have designed a small geographic information system to analyze flooded under sea-level rise of 1 m in mainland China, and it for EMRICES simulate provides a good test. GIS analysis obtained without reducing emissions, rising sea levels will make China's coastal areas (excluding South China Sea islands) will submerge 30,000 sq km, the simulation results are almost consistent with EMRICES. GIS analysis shows that China Bohai Sea, Lixiahe watershed, Shandong coastal, Tianjin some areas along the coast, Qinhuangdao and Pearl River Delta region have a larger area to be flooded.

The study also gives us some policy implications: First, sea level rise will bring a lot of physical effects, including loss of land and wetlands, etc., which impact on China's living in the world, worse than the United States; secondly, thus will bring some direct economic losses, but the impact is not great relative to GDP, but this is not considered an indirect impact on the economy; third, adaptation measures to reduce the loss of protection has played a significant role; in China, has protected and unprotected contrast situations differ loss of land area has eight times the economic losses resulting difference is also more than four times. Therefore, China needs to actively participate in climate protection in addition, there are local conditions necessary to develop appropriate protective measures to cope with rising sea levels submerged hazards.

References

Darwin T (2001) Estimates of the economic effects of sea level rise. Environ Resour Econ 19: 113–129

David A, Tol RSJ (2012) The climate framework for uncertainty, negotiation and distribution (fund), technical description, version 3.6

Dennis KC, NiangDiop I, Nicholls RJ (1995) Sea level rise and senegal: potential impacts andconsequences. J Coast Res, Special issue 14:243–261

Gambarelli G, Goria A (2004) Economic evaluation of climate change impacts and adaptation in Italy. FEEMN di L 103:04

Hoozemans FMJ, Marchand M, Pennekamp HA (1993) A global vulnerability analysis: vulnerability assessment for population, coastal wetlands and rice production and a globalscale (second, revised edition). Delft, Delft hydraulics

IPCC (2006) The 2006 IPCC guidelines for national greenhouse gas inventories (2006 guidelines). http://www.ipcc-nggip.iges.or.jp/public/2006gl/index.html, http://www.news.sciencenet.cn/htmlnews/2014/3/289997.shtm

Joshi MM, Turner AG, Hope C (2013) The use of the land-sea warming contrast under climate change to improve impact metrics. Clim Change 117:951–960

Nicholls RJ, Tol RSJ, Vafeidis A (2008) Global estimates of the impact of a collapse of the West Antarctic ice sheet: an application of FUND. Clim Change 91(1):171–191

Nordhaus W (2007) Critical assumptions in the Stern Review on climate change. Science 317: 201–202

Sørensen B (2008) Pathways to climate stabilization. Energy Policy 36:3505–3509

Stem N (2006) Stem review on the economics of climate change. Report to the Prime Minister and the Chancellor of the exchequer on the economics of climate change

UNDP (2008) Fighting climate change: human solidarity in a divided world. Human development report 2007/2008. http://hdr.undp.org/en/reports/global/hdr2007-2008/

Vermeer M, Rahmstorf S (2009) Global sea level linked to global temperature. PNAS 106 (51):21527–21532

Wang Z, Li HQ, Wu J et al (2010) Policy modeling on the GDP spillovers of carbon abatement polices between China and the United States. Econ Model 20:40–45

Wang Z, Zhang S, Wu J (2012a) A new RICEs model with the global emission reduction schemes. Chin Sci Bull 57(33):4373–4380

Wang Z, Wu J, Zhu QT et al (2012b) MRICES: a new model for emissions mitigation scheme assessment and its application. J Geogr Sci 22(6):1131–1148

Zeider RB (1997) Climate change vulnerability and responses strategies for the coastal zones of Poland. ClCh 36:151–173

Chapter 4
EMRICES

4.1 Introduction

At present, countries around the world have realized that they need to tackle climate change together, and negotiations coping with this issue had been held. Therefore, the games on carbon mitigation among countries were brought in. The basic motivation for the world to achieve the agreement of carbon mitigation scheme is that every country could benefit from the scheme or even to maximize their interests. However, due to the development bias among those countries, fairness on carbon mitigation has become the primary requirement. All these pursuits require the completion of a global game analysis. As a result, a research on global carbon mitigation cooperation, from the perspective of game theory, is of great significance.

Currently, many scholars have researched on global greenhouse gas mitigation in the perspective of game theory. Researches in this field can be divided into two categories: one kind of the models focus on local problem in the global climate protection. Such as Scheffran and Pickl (2000) discussed the possibility of international cooperation in joint implementation mechanism. Babiker (2001) used CGE model to talk about the attachments of Kyoto Protocol the possibility of countries' promises to reduce carbon emissions. Caparros et al. (2004) researched the carbon mitigation game issues between developed and underdeveloped countries under incomplete information. Kemfert et al. (2004) discussed the national climate mitigation cooperation under the influence of the international trade. Haurie et al. (2006) constructed a differential game model based on carbon emissions trading system. Tavoni et al. (2011) simulated the mitigation cooperation among the crowd of different disposition in uncertain environment in experimental economics. The other kind of models focuses on the achieving of global mitigation scheme. Such as Tavoni et al. (1996), they divided the world into five areas, and launched game strategy case studies. Yang and Sirianni (2010) used the RICE model to research the non-cooperative Nash equilibrium, and discussed the feasibility of carbon quota allocation according to the principle of "color preservation". Some of the models

Z. Wang et al., *Integrated Assessment Models of Climate Change Economics*, DOI 10.1007/978-981-10-3945-4_4

the studies used are based on IAM (integrated assessment model), such as Kemfert et al. (2004), Yang and Sirianni (2010) their works are based on WAGEM model and RICE model, respectively. These models are described in detail on climate change and economic development. Some models in other research simplified these equations, and put more emphasize on game mechanism. However, these models have some common flaws. First, these models set game strategy mainly according to the rate of reduction in base scenario rather than from the global carbon emissions control, but the real effective scheme to achieve according to IPCC mitigation aims is to control the total emissions. Second, these models seldom considered problems from the perspective of climate ethics. Wang et al. (2014) believe that the Pareto improvement of mitigation scheme may be the fundamental guarantee to set an accordant mitigation scheme worldwide. In this chapter, on the basis of the national climate change ethics,we developed the MRICE model, and by using which, we simulated and analyzed the economic growth and carbon emission of all the countries or regions around the world and what the world would like to do in the framework of game theory. MRICES model has great advantage in IAM. Firstly, the method we adopt is IAM, which named MRICES. We improved a great flaw of most of IAM in economic module. IAM usually did not consider the economic relationship among countries. We add GDP-spillover module to overcome the flaw. And the endogenous technology improvement module is also an innovation in IAM. Therefore, we have novel insights in the method. Secondly, considering the economic connection among countries, the climate games will be quite different. We have also done the parameter sensitivity analysis in our paper. We discussed how discount rate will affect the game results and the Pareto-consistent solution. We think these work is important.

As the public goods characteristic of carbon dioxide, reducing emissions is actually a game of non-cooperative Nash equilibrium without the global emissions constraint. It means the reduction strategies of all countries will affect the welfare correspondingly. At the same time, countries would choose the best mitigation scheme for maximizing their welfare. Therefore, the standard used to determine whether each country could benefit from the reduction scheme is that if it could bring greater benefit for each country than Nash equilibrium could. The purpose of this chapter is to seek a global carbon mitigation scheme so as to match the principle of Pareto improvement.

4.2 Analysis Framework

4.2.1 The Model

Climate change will bring loss to global economy. Global mitigation scheme aiming at climate change will reduce the loss, and countries would benefit from the mitigation result. But participating in mitigation needs to limit current economic development, and would bring corresponding costs. Measuring the loss and benefits is the key to the model. IAM model is such an international popular model, which combine the economic module and climate module together to evaluate the economic loss and the cost of carbon mitigation scheme (Lessmann et al. 2009). The IAM model in this research is MRICES (Multi-factor Regional Climate and Economy System) (Wang et al. 2012), which is developed based on RICE model initially developed by Nordhaus and Yang (1996). In the past decades, RICE model played a pivotal role in the assessment of climate change, which can be found in the 2nd, 3rd, 4th IPCC assessment reports. MRICES divides the world into eight areas, and it retains the original advantages of RICE. MRICES treats climate change factors as production factors that affect economy. In particular, by means of the GDP spillover modules, MRICES model shows the mutual influence of the global economy. It is suitable for the research of national economic externality when reducing the carbon emissions and it is better to reflect the mitigation policies. The MRICES also take endogenous technology progress into account simultaneously, which means the technology progress depends on the accumulation of R&D capital (Wang et al. 2012). These improvements have strengthened the connection of regional economies, and also reflect the essence of technological progress. The MRICES is currently used to assess some mitigation schemes (Wang et al. 2012). This chapter will analyze the cooperative game behavior of all countries on mitigation based on EMRICES.

4.2.2 The Situation of Global Carbon Mitigation

IPCC report has proposed the control target of global climate change that global temperature should not increase by 2 °C till 2100, this is the consensus at Copenhagen 2008. It is generally considered that reducing global carbon emission is inevitably required to fulfill this control target (Nordhaus and Boyer 2000; IPCC 2007; Stern 2007). Then, how much emissions is left in the future if global temperature increased no more than 2 °C till the year 2100? By using climate module of MRICES, This research assumes that global carbon emissions change linearly every year, the simulation result shows that if the global temperature increased no more than 2 °C by 2100, the maximum global carbon emissions in 2100 would be 3.18 GtC. Furthermore, in this model the global carbon emissions in 2010 was 8.2 GtC as the beginning point,which means in 2100, the global carbon emissions should be reduced to 38.8% of the emissions of 2010. Moreover, the per capita CO_2 emissions of Middle-income countries in 2010 was 3.46 GtC, approximately equaled to 70.9% of the world's average per capita CO_2 emissions (4.88 GtC). It means that once the 2 °C limit was considered, the vast majority of countries including middle income countries would confront a total emission reduction even from the perspective of equal per capita CO_2 emissions, otherwise the targets would be too difficult to achieve. Therefore, the target of controlling within 2 °C inevitably requires the global participation in mitigating. In this chapter, we will study how to reach a global agreement in the view of game theory with global participation to total CO_2 emissions reduction.

4.2.3 Global Mitigation Principles

In order to understand the basis of the game, we need to discuss some ethical principles about the global cooperation to reduce emissions. There are two principles for the global cooperation mitigation schemes considering reduction target. The first one is the final solution should be accepted by each country or region. The second one is, in the perspective of fairness, we should maintain growth opportunities for developing countries as much as possible, because the developed countries have historical responsibility on the issue of global warming. The mitigation scheme that can be accepted should possess following character that every country and region would benefit from it, satisfying the principle of Pareto improvement. In other words, the interests of all the participants should not be damaged because carbon dioxide is the world's largest public goods. In the perspective of welfare economics, the purpose of mitigation is to improve the welfare of human beings, so it is to a specific nation. The motion of one country to reduce CO_2 emissions depends on whether it could benefit from the action.

Basing on the principles above, a variety of programs could be designed to meet the principle of Pareto improvement. However, the goal of this chapter is to take developing countries interests into consideration for the fairness.

To sum up, in this chapter, the principle of designing mitigation scheme has three points. Firstly, meet the emission reduction requirements. In specific, global temperature should not exceed 2 °C in 2100. Secondly, match the principle of Pareto improvement. Thirdly, keep fairness as much as possible.

4.3 The Game Design and Simulation

4.3.1 Welfare

The criteria that determined the willingness of one country or region whether cooperate with each other to reduce carbon emissions should be defined at the first time. In economics, welfare is usually seemed as the pursuit of biggest interest. The welfare is generally defined as a function of consumption. For simplicity, $y = U(C)$ denotes welfare, C means consumption. Thus, $U(C)$ meets $\frac{\partial U}{\partial C} > 0, \frac{\partial U^2}{\partial^2 C} < 0$. This article also follows this principle. Specific form is $U(C) = \frac{(C/L)^{1-\tau}}{1-\tau}, \tau < 1$, where, τ is risk-averse coefficient. The larger τ is, the less obvious the increase of welfare caused by increased consumption is. And the variable L is population.

The benefits of mitigation will reflect the total incomes in the future for a long period. In fact, the benefit of each country will decline at the beginning of mitigation, and the incomes of climate changing brought by mitigation will be well represented in the long run. Therefore, this chapter defines the welfare value as the cumulative welfare value starting from the beginning year to the year 2100. Furthermore, the future welfare is the NPV (the net present value) calculated by the discount rate ρ. At the same time, this chapter accordingly calculated the total welfare of total population in each country or region. The specific form is:

$$U_i(C_{i,t}, L_{i,t}) = \sum_{t=1}^{n} (1+\rho)^{-t} L_{i,t} \frac{\left(C_{i,t}/L_{i,t}\right)^{1-\tau}}{1-\tau}$$

4.3.2 *The Mitigation Strategy*

As discussed above, to achieve the global climate change control objective that the temperature could not exceed 2 °C till 2100, all the countries including middle developed countries should reduce the total emission. Thus, we set the game strategy as reduction rate of the total amount of each country or region in 2050 relative to benchmark year; and after 2050, all countries or regions would keep the reduction rate as 2050 till 2100. The division of world and the corresponding benchmark year are shown in Table 4.1.

It assumes that the game G has eight participants which are the eight countries or regions in MRICES. Each participant has a strategy space: $S_1, S_2, \ldots, S_8$, and $s_{ij} \in S_i$ means the strategy j of the game participant. $j \in \{1, 2, \ldots, 11\}$; $S_i = \{0, 0.1, 0.2, \ldots, 0.9, 1\}$. If $s_{i2} = 0.1$, it means the strategy 2 of the game participant *i* is: till the year 2050, the total carbon emission of a country or region reduces by 10% than the benchmark year. The payoff function is U_i, means the NPV of the cumulative welfare values of the country or region *i* to the year 2100. This game can be expressed as $G = \{S_1, \ldots, S_8; u_1, \ldots, u_8\}$.

It is hard to use the traditional two-dimensional table to show game structure because there are 8 main participants in the game. In this chapter, in order to illustrate game structure vividly, game-agent are simplified to two parts: a country (such as EU) and other countries,and then explained through the two-dimensional table. As shown in Table 4.2, The EU and other countries are game participants, the mitigation rate is game strategy, and Game income is the accumulated value of the participants in game.

Table 4.1 Benchmark year of each country or region

Country or region	Benchmark year	Country or region	Benchmark year
China	2005	High developing countries	2005
America	1990	Middle developing countries	2005
Japan	1990	Low developing countries	2005
EU	1990	Developed country countries	1990

Table 4.2 The structure of the game

		EU			
Other country	Abatement rate	0	0.1	0.2	0.9
	0	(welfare of EU, welfare of other country)			
	0.1				
	0.2				
	0.9				

4.3.3 The Solution of the Nash Equilibrium

It's hard to find the Nash equilibrium of the game from the analytical view because of the complexity of MRICES. So we need to redesign the solution algorithm. This chapter design algorithm as follows: firstly, preset any set of strategies $\left(s_1^0, s_2^0, \ldots, s_8^0\right)$ as initial solution, and calculate the best mitigation scheme s_{1j^*} from $\left(s_{ij}, s_2^0, \ldots, s_8^0\right)$, and make the first country or region's welfare U_1 largest. Next, find out the best mitigation scheme for the other countries successively. The calculation will stop if there is not any change about strategy for all the countries, which means $\left(s_1^1, s_2^1, \ldots, s_8^1\right) = \left(s_1^0, s_2^0, \ldots, s_8^0\right)$. Otherwise, start a new round of calculation until $\left(s_1^k, s_2^k, \ldots, s_8^k\right) = \left(s_1^{k-1}, s_2^{k-1}, \ldots, s_8^{k-1}\right)$. Then, the Nash equilibrium solution is $\left(s_1^*, s_2^*, \ldots, s_n^*\right) = \left(s_1^k, s_2^k, \ldots, s_8^k\right)$. Calculating flow is shown the Fig. 4.1.

This chapter use MRICES as platform and take computation algorithm in Fig. 4.1. Calculate Nash equilibrium solution with the initial value $(0, 0, 0, 0, 0, 0, 0, 0)$, $(0.1, 0.1, 0.1, 0.1, 0.1, 0.1, 0.1, 0.1)$, $(0.2, 0.2, 0.2, 0.2, 0.2, 0.2, 0.2, 0.2)$, …, $(1, 1, 1, 1, 1, 1, 1, 1)$. After calculation, the final solution is $(0, 0, 0, 0, 0, 0, 0, 0)$ under the given initial value. That is to say, the Nash equilibrium solution we found is that none of the countries or regions will adopt the mitigation strategy.

In order to prevent missing any solution, this chapter increase every country's strategy set $S_i = \{0, 0.05, 0.1, \ldots, 0.95, 1\}$, and the final result is still $(0, 0, 0, 0, 0, 0, 0, 0)$.

The result of game shows that, if each country or region only cares about its own maximum benefit, it would lead that no one would take action in mitigation. However, cooperation can improve each country or region's welfare. Therefore, the global will be trapped in prisoner's dilemma if the problem can not be solved through negotiation on the issue of climate protection.

4.3.4 The Mitigation Scheme

This chapter searched for effective Pareto scheme based on MRICES and found out that a scheme could meet the needs. The scheme is: China should reduce 20% carbon emissions on 2005 levels by 2050; USA should reduce 63% carbon emissions on 1990 levels by 2050; Japan should reduce 85% carbon emissions on 1990

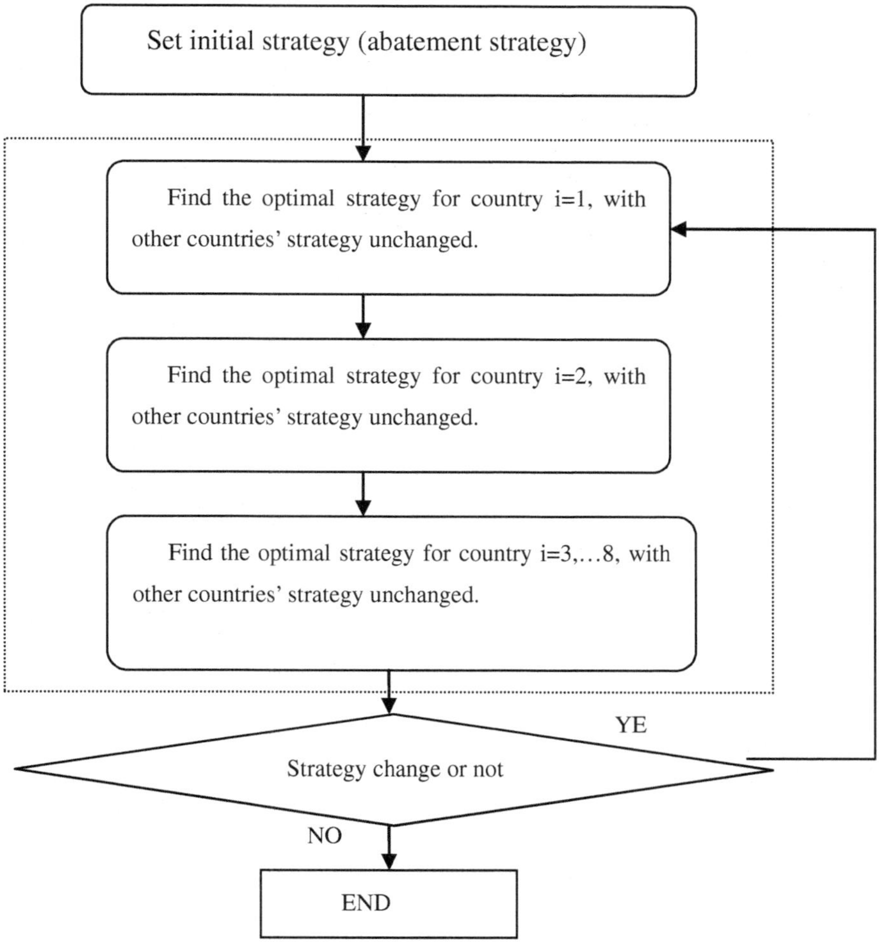

Fig. 4.1 The algorithm of the solution to the Nash equilibrium

levels by 2050; EU should reduce 85% carbon emissions on 1990 levels by 2050; High developed countries should reduce 50% carbon emissions on 2005 levels by 2050; Middle developed countries should reduce 20% carbon emissions on 2005 levels by 2050; Developed countries should reduce 82% carbon emissions on 1990 levels by 2050. In the scheme, the global temperature increases by 1.996 °C in 2100. Therefore, the global temperature increment can be controlled within 2 °C, and meet the moral constraints. Figure 4.2 shows the change trends of atmospheric temperature and ocean temperature under this scheme.

It should be presented that, when the US reduction rate is too high, the cumulative benefits could not be improved until 2100 due to the higher value of current benefits in United States. China's emissions reduction rate is lower than that of high developing countries mainly because of China's role in the world economy and

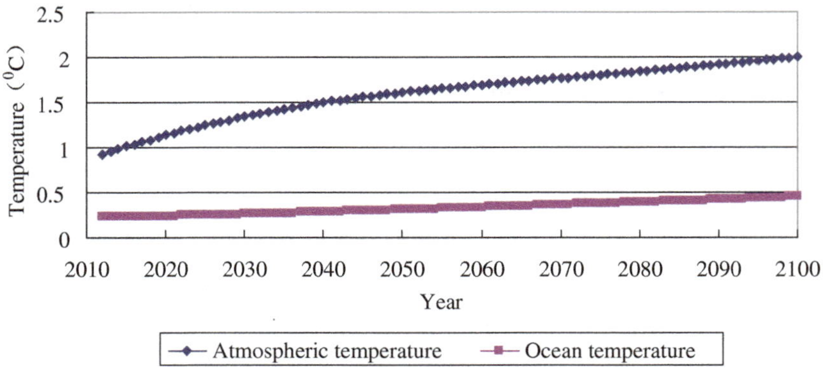

Fig. 4.2 The temperature change trend chart of Pareto scheme

China's current low level of per capita carbon emissions. If China's emissions rate was too high, China's per capita carbon emissions levels would not be improved. The world economy would be bigger affected since China and USA are the engines of world economy, with financial and manufacturing sectors, respectively.

From the perspective of the cumulative welfare changes shown in Figs. 4.3 and 4.4, the cumulative welfare of China is 1.84% higher than that in the condition of non-cooperative Nash equilibrium in 2100. The accumulation of USA's welfare is 0.01% higher; the accumulation of Japan is 0.38% higher; the accumulation of European is 0.19% higher; the accumulation of high developing countries is 0.88% higher; the accumulation of middle developing countries is 1.66% higher; the accumulation of low developing countries is 3.21% higher; the accumulation of developed countries is 0.001% higher. The welfare values from all countries or regions are better than that of the non-cooperative Nash equilibrium, so the scheme meet the Pareto improve principle.

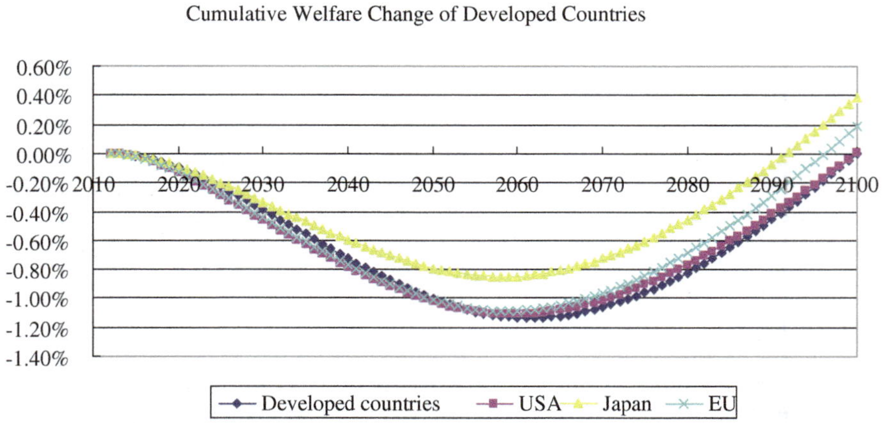

Fig. 4.3 Developed countries' accumulate welfare change trend

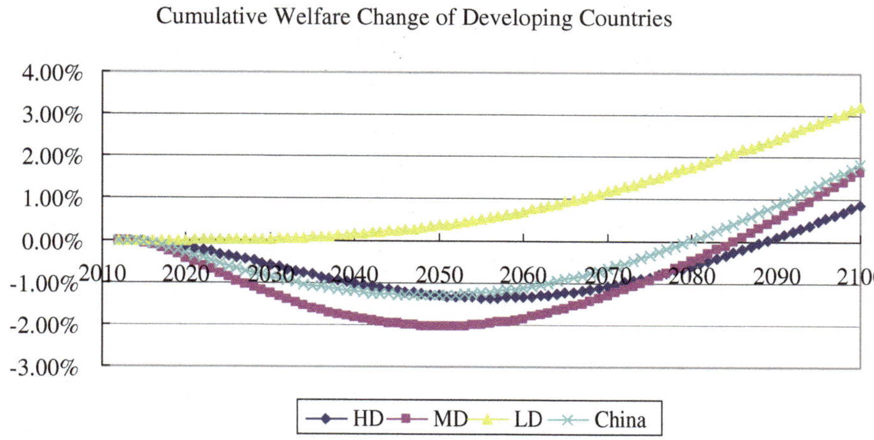

Fig. 4.4 Cumulative welfare change rate in developing countries

In brief, the results shows, the cooperation scheme is better than none of the countries adopt mitigation strategy. From the ethics, it is not only the effective way to solve prisoner's dilemma, but also reasonable.

4.4 Sensitivity Analysis

There are many parameters in the MRECIS model, and many scholars have discussed about the parameters in IAM, especially about the analysis of the discount rate in welfare. Nordhaus' discount rate is 1.5% (per year), Stern's discount rate is 0.1% (per year). The differences of the discount rate also caused many scholars' debate (Mendelsohn 2006; Stern 2006; Nordhaus 2007; Weitzman 2007). Discount rate reflects to which extent people care about the future. The higher the discount rate is, the less people would care about the future welfare, and vice verse. Beckerman and Hepburn 2007 raised this issue of discount rate to the ethic level, considering that the discount rate is the important parameter to reflect intergenerational equity. So the sensitivity analysis to the discount rate was done in this chapter.

4.4.1 The Nash Equilibrium

Low discount rate means that future losses would be high when it discounts to present. So it may affect the emission reduction strategy. By changing the discount rate in MRICES and calculating at 0.5% intervals from 0.01% to 3%, we can find Nash equilibrium is that all countries do not take the total mitigation strategy. The discount rate does not affect the Nash equilibrium within the scope of the discount rate. The policy implication of the result is that the cumulative welfare values would be very different under the different discount rates. But if all countries try to maximize their own welfare without cooperation, they still don't take action in mitigation.

4.4.2 *The Pareto Principle*

While Nash equilibrium is not affected by the discount rate, different discount rate can be further investigated that whether the Pareto improvement of mitigation scheme could be satisfied. So we calculated when discount rate equals 1%, 1.5% and 3%. The previous result is when the discount rate is 1.5% in this chapter.

In Stern scenario, the discount rate is 0.1%, the accumulative welfare of each country increases higher compared to 1.5% discount rate. The cumulative welfare improvement in developing countries is still greater than that of the developed countries. The mitigation scheme satisfies the principle of Pareto improvement (Figs. 4.5 and 4.6).

In order to observe the effect of the discount rate on Pareto improvement, the scenario of the discount rate equals 3% was calculated. In this scenario, the accumulation welfare of countries will become smaller compared with Nordhaus scenario, and the accumulation of welfare can be improved in developing countries, while the cumulative benefits of developed countries is not improved. This scheme cannot satisfy Pareto improvement (Figs. 4.7 and 4.8).

Simulation results showed that when the discount rate increases, the original scheme of the nature of the Pareto improvement may be broken, and range of

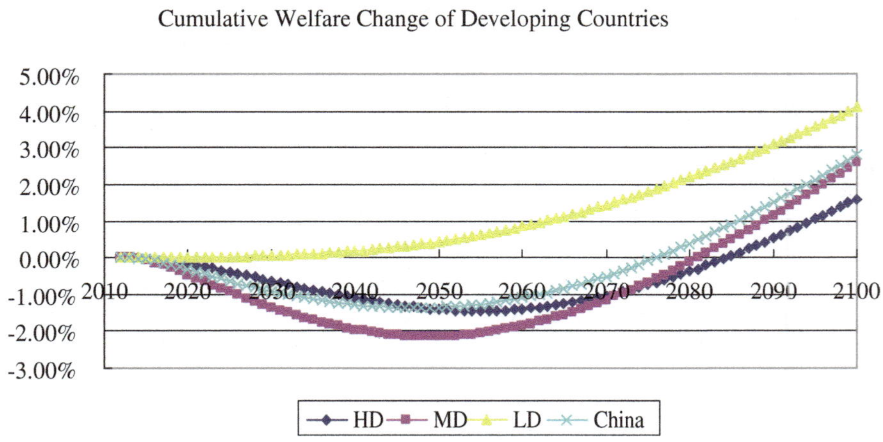

Fig. 4.5 Developed countries' accumulate welfare change trend when the discount rate is 0.1%. (Notation: HD represents for high developing countries, MD represents for middle developing countries, LD represents for low developing countries)

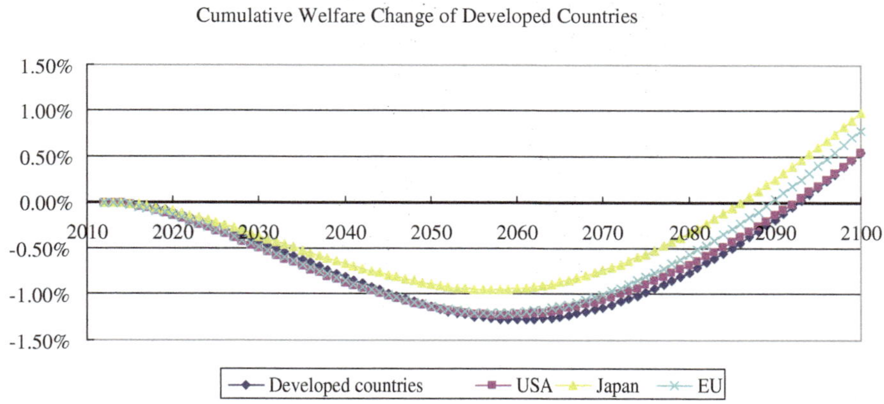

Fig. 4.6 Developing countries' accumulate welfare change trend when discount rate is 0.1%

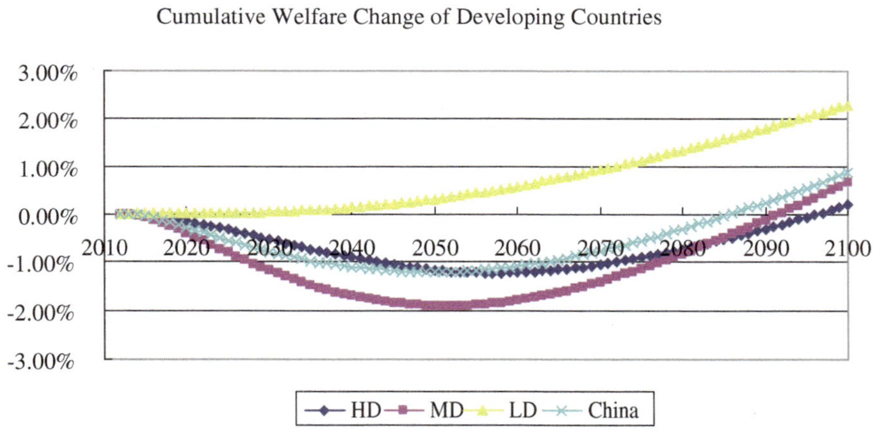

Fig. 4.7 Developed countries' accumulate welfare change trend when discount rate equals 3%

schemes which meet Pareto improvement would narrow down. When the discount rate decreases, the original scheme of the nature of the Pareto improvement may not be broken, and range of schemes which meet Pareto improvement will enlarge. This is mainly because when the discount rate is higher, people care less about the welfare of the future. The welfare change characteristic of mitigation is that the welfare will be lower at the early stage because of mitigation, but at the later stage, the welfare will be higher because the decreased temperature reduces the economic losses.

An obvious conclusion is that the higher the discount rate of a country is, the harder the cumulative benefits could be improved. Therefore, the less likely the country would participate in the scheme. There may be differences between the discount rate in different countries (Davidson 2006). Developing countries are

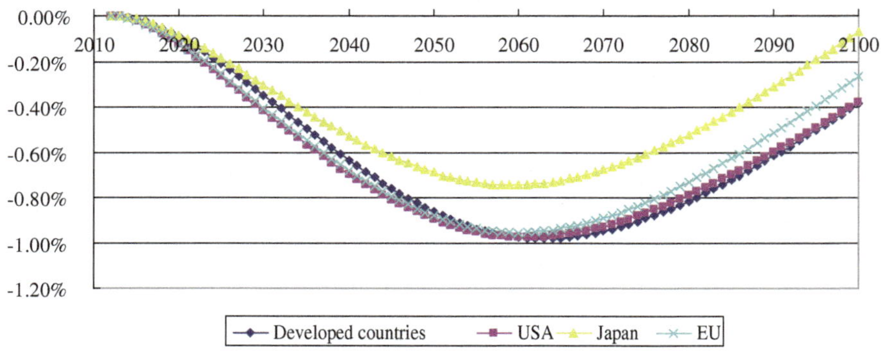

Fig. 4.8 Developing countries' accumulate welfare change trend when discount rate equals 3%

under high-speed infrastructure renewal and construction at present, while developed countries infrastructure renewal has been basically finished. Thus, Developing countries' discount rate may be higher than the developed countries'. Consider this situation, developing countries' task of mitigation may be lower than the reduction rate involved in this chapter.

4.5 The Conclusion

This research constructs three global mitigation cooperation plans in the perspective of global mitigation cooperation scheme. Firstly, keep the agreement of global climate change control, which means that meet the Copenhagen Consensus controlling global temperature would not raise over 2 °C by 2100. Secondly, meet the principle of Pareto improvement, and all countries' cumulative welfare can benefit from this mitigation. Thirdly, reflect the fairness of the principle of "common but differentiated responsibilities". On this basis, this study proposes a scheme which can meet the three principles, and provide theoretical support for further climate negotiations.

(1) Considering some important parameters, such as discount rate, this chapter examines the influence of discount rate on non-cooperative game equilibrium and the impact on the characteristics of the principle of Pareto improvement of the mitigation scheme.
(2) The results revealed that the discount rate does not affect the Nash equilibrium. It means no matter how important people valued future, the importance of future is not more than current's, because the discount rate is positive, if all countries only care for their own benefit maximization, no one will take the total mitigation strategy.

(3) The discount rate will affect the characteristics of Pareto improvement. In general, the lower the discount rate is, the more the schemes meet the characteristic of Pareto improvement, and vice versa. In particular, if a scheme meets Pareto improvement, as long as the discount rate is smaller than the value of the original scheme adopted, the scheme would meet the characteristics of the Pareto improvement; If the discount rate is bigger than the original one, the scheme would not necessarily meet the characteristics of the Pareto improvement, since the scheme proposed in this chapter meets Pareto improvement. But when the discount rate is 3%, it does not meet the Pareto improvement scheme. When adopted the discount rate of 0.1%, the result not only meets the Pareto improvement scheme, but also shows more space for improvement.

Of particular note is,because of the difficulty of data acquisition, the model in this chapter have adopted the same discount rate at all the countries; but the discount rate of developing countries is much higher than that in developed countries since developing countries are in a fast development period of infrastructure renewal and construction. If taking this fact into consideration, the mitigation scheme designed in this research still need further revision, which means the reduction rate of developing countries should be lower.

Acknowledgements We are grateful for the financial support which was provided by Grant No. 2012CB955804 from the National Basic Research Program of China (973 Program).

References

Babiker MH (2001) The CO2 abatement game: costs, incentives, and the enforceability of a sub-global coalition. J. Econ. Dyn. Control 25(1–2):1–34

Beckerman W, Hepburn C (2007) Ethics of the discount rate in the Stern Review on the economics of climate change. World Economics-Henley on Thames 8(1):187

Caparros A, Pereau JC, Tazdait T (2004) North-South climate change negotiations: A sequential game with asymmetric information. Public Choice 121(3–4):455–480

Davidson MD (2006) A social discount rate for climate damage to future generations based on regulatory law. Clim Change 76(1–2):55–72

Fankhauser S, Kverndokk S (1996) The global warming game—simulations of a CO2-reduction agreement. Resour Energy Econ 18(1):83–102

Haurie A, Moresino F, Viguier L (2006) A two-level differential game of international emissions trading. Birkhauser Boston, Cambridge

IPCC (2007) Clim Change. http://www.ipcc.ch

Kemfert C, Lise W, Tol RSJ (2004) Games of climate change with international trade. Environ Resour Econ 28(2):209–232

Lessmann K, Marschinski R, Edenhofer O (2009) The effects of tariffs on coalition formation in a dynamic global warming game. Econ Model 26(3):641–649

Mendelsohn RO (2006) Critique of the Stern Report, A. Regulation 29:42

Nordhaus W (2007) Economics. Critical assumptions in the Stern Review on climate change. Science (New York, NY) 317(5835): 201

Nordhaus WD (2007) A review of the Stern Review on the economics of climate change. J Econ Lit 686–702

Nordhaus W, Boyer J (2000) Warming the world: the economics of the greenhouse effect. MIT Press, Cambridge, MA

Nordhaus WD, Yang Z (1996) A regional dynamic general-equilibrium model of alternative climate-change strategies. Am Econ Rev 741–765

Scheffran J, Pickl S (2000) Control and game-theoretic assessment of climate change: options for Joint Implementation. Ann Oper Res 97:203–212

Stern N (2006) Review on the economics of climate change. London HM Treasury

Stern N (2007) The economics of climate change: the Stern review. London, cambridge University press

Tavoni A, Dannenberg A, Kallis G, Loschel A (2011) Inequality, communication, and the avoidance of disastrous climate change in a public goods game. Proc Natl Acad Sci USA 108(29): 11825–11829

Wang Z, Liu X, Liu CX, et al (2014) Several issues of climate change ethics (in Chinese). Chin Sci Earth Sci 44:1600–1608

Wang Z, Wu J, Zhu Q, Wang L, Gong Y, Li H (2012) MRICES: a new model for emission mitigation strategy assessment and its application. J Geogr Sci 22(6):1131–1148

Wang Z, Zhang S, Wu J A new RICEs model with the global emission reduction schemes. Chin Sci Bull (26): 2507–2515

Weitzman ML (2007) A review of the Stern review on the economics of climate change. J Econ Lit 45(3):703–724

Yang Z, Sirianni P (2010) Balancing contemporary fairness and historical justice: a ‘quasi-equitable’ proposal for GHG mitigations. Energy Econ 32(5):1121–1130

Chapter 5
The Analysis for Synergistic Effect of Policy of Environmental Tax with Dynamic CGE in China

5.1 Introduction

Environmental tax, also known as eco taxes, green taxes, is to solve the environmental pollution, which rose in international taxation academia at the end of the twentieth century; there is not a clear definition yet. In general, the environment tax internalizes the social cost of ecological destruction and environmental pollution into the production cost, the producers' and the consumers' behaviors are affected through the market price, the distribution of environmental resources is decided by market mechanism. Environmental tax has a relatively perfect system and tax experience in Europe and the United States. In China, current levied taxes which are related to environmental protection include resource tax, travel tax, urban maintenance and construction tax, vehicle purchase tax, etc. These taxes have played a certain role on China's resources and environment protection, but they are not specifically according to the resource protection. Environmental tax hasn't been levied yet. With the rapid development of China's economy and the acceleration of the industrialization process, demand and exploitation of resources has also increased sharply, coupled with China's rapid population growth and urbanization level enhancement unceasingly, making the poor resource situation worse. In order to improve the increasingly severe situation of resources and environment, the demand for environment tax is becoming more and more urgent.

For a long time, China's energy structure dominated by coal, causing serious air pollution such as sulfur dioxide and soot, endanger the health of residents. The economic loss caused by acid rain is even more astonishing. In 1992, China began to carry out the sewage charges work on SO_2 emission into provinces and nine municipalities. In 1998, SO_2 charging work was expanded to control area of SO_2 and control area of acid rain (two control zones). With the collection of sewage charges, a lot of problems exposed, such as the narrow taxation collection, low charging standard, it is difficult to play the effect of protecting environment and saving resource. In order to solve the increasingly serious environmental problems,

Z. Wang et al., *Integrated Assessment Models of Climate Change Economics*, DOI 10.1007/978-981-10-3945-4_5

the sulfur tax is considered to be an effective policy means proposed by many scholars. Wu and Xuan (2002) used the CGE model to analyze reduction effects and economic effects of sulfur tax. They found that the sulfur tax can significantly reduce SO_2 emissions and improve the quality of the environment, and they estimated the optimal tax rate of the sulfur is 400–660 yuan/ton SO_2 (Wu and Xuan 2002). Based on China's input-output data 2002, Ma (2008) simulated the effects of sulfur tax by CGE model, the results showed that the sulfur tax can reduce SO_2 emission and would not produce a big negative impact on GDP. The energy structure can also be improved, the optimal tax rate of the sulfur is 460–1050 yuan per ton of sulfur (Ma 2008). Using directional environmental production frontier function model, Tu (2009) estimated the shadow prices of China's provincial industrial SO_2 emissions. It was found that shadow price of industrial SO_2 emissions have a great relationship with the discharge scale, the shadow price of industrial SO_2 emission is different in different provinces, the shadow price of industrial SO_2 discharge in Beijing area is 2800 million yuan per ton, the national average is 200 million yuan per ton (Tu 2009). Using the directional distance function model, Yuan and Cheng (2011) estimated the shadow price of SO_2 emissions from the industrial sector at China's 284 prefectural-level cities, the results showed that the shadow price of industrial SO_2 from 2003 to 2008 was 51.58 thousand yuan per ton (Yuan and Cheng 2011).

CO_2 is considered to be the most important greenhouse gas, greenhouse gas emissions control in recent years mainly revolves around to reduce CO_2 emissions. China is a big energy consumption and emission power, due to the development of the economy and urbanization speeding up, energy demand will rise further. With increasing CO_2 emissions in the atmosphere, the carbon tax is considered to be one of the effective means to control global warming. However, to reduce CO_2 emissions and the use of fossil fuels, is bound to affect the economic development of China. Under the background of global economic integration, it will also affect to the world economy through international trade. Therefore, before the implementation of emission reduction policy, it is necessary to simulate the economic impact of reductions in carbon emissions and emissions reduction effect.

In fact, many scholars in China as early as 2002 had already started to analyze carbon emission reduction effect and economic effect of levying a tax on carbon. He et al. (2002) used the static CGE model to simulate the impact of carbon tax on the national economy under the different targets, and estimated the marginal cost of CO_2 which is between 88.4 and 418.2 Yuan/ton, the equivalent of 324.1–1533.4 Yuan/ton carbon (He et al. 2002). Gao and Chen (2002) adopted MARKAL MACRO model and set up different carbon tax levels, such as 30 dollars per ton of carbon, 50 dollars per ton of carbon, 100 dollars per ton of carbon, 150 dollars per ton of carbon and 200 dollars per ton of carbon, the results showed that higher rates would produce a larger negative economic effects, emission reduction effect is not ideal. When the tax rate is 50 dollars per ton of carbon, carbon emissions effect is the best, and the economic consequences is the smallest, the equivalent is 413 Yuan per ton of carbon with 2002 exchange rates (Gao and Chen 2002). Wang et al. (2005) simulated macroeconomic effects under different reduction rate using dynamic recursive CGE model, the results showed that with the increase of reduction rate, the marginal cost of carbon emissions also increases accordingly. Under the reduction rate of 10–30%, carbon abatement costs are 100 Yuan per ton and 470 Yuan per ton, respectively (Wang et al. 2005). Cao (2009), using a recursive dynamic CGE model, simulated the economic and environmental impact of carbon tax, the study found that a carbon tax (50–200 Yuan per ton of carbon) have less effect on the economy, but the carbon reduction effect is remarkable (Cao 2009). Wang et al. (2009) adopted the comprehensive evaluation model of China's energy policy—energy economy model (IPAC–SGM) developed by the national development and reform commission to simulate the impacts of different carbon tax on China's CO_2 emissions and macro economy, the tax rate of high, medium and low solution respectively are 100–200 Yuan per ton of carbon, 50–100 Yuan per ton of carbon, 20–40 Yuan per ton of carbon, in order to avoid the negative economic effects of carbon tax, China's carbon tax should start with lower rate, then gradually improve (Wang et al. 2009). Zhu et al. (2010), based on CGE model, set up high rate (100 Yuan per ton of carbon), medium rate (50 Yuan per ton of carbon), and low rate (20 Yuan per ton of carbon), respectively. The reduction effect and economic effect of carbon as production tax and consumption tax are analyzed; the results found that carbon tax can reduce CO_2 emissions, the reduction effect of carbon as production tax is better than carbon as consumption tax (Zhu et al. 2010). Li (2014) by using the static CGE model, set up five different carbon tax, 10 Yuan per ton of carbon, 20 Yuan per ton of carbon, 40 Yuan per ton of carbon, 180 Yuan per ton of carbon and 300 Yuan per ton of carbon, respectively. The results found that carbon tax can reduce CO_2 emissions and adjust energy structure, but it will have certain influence on the economy, so carbon tax should started with lower rate (Li 2014).

As the voice of environmental taxes growing in recent years, the environment tax rate design becomes the focus in the academic circles. During the research above only one environmental tax is analyzed. What effect does levying environment tax have? What kind of impact will it have on the economy? Will different types of taxes have influence on each other? How to design tax rates to protect the environment at the same time not causing serious economic fluctuations? To answer these questions, this article, taking sulfur tax and carbon tax as examples, simulated the impact of environment tax on the macro economy and emission reduction effect based on the dynamic general equilibrium model.

5.2 Model and Data

5.2.1 CGE Dynamic Mechanism

This article mainly is to simulate the impact on the macro economy after levying environment tax, thus the CGE equation system is simplified, detailed CGE equation system can be seen in Wang etc. Wang et al. (2010). CGE dynamic equation is given here. CGE dynamic adopted Solow growth model, the source of economic growth is the dynamic of capital, the capital is equal to the capital stock of the previous period to deduct depreciation plus new investment (Liu et al. 2013).

$$K_{t+1} = (1 - \delta_k)K_t + I_t \tag{5.1}$$

Capital growth rate equation is deduced.

$$\frac{\dot{K}}{K} = \frac{I_t - \delta_k K_t}{K_t} \tag{5.2}$$

$\dot{K}$ denotes the increase of capital. In addition, the production technology level and labor are dynamic.

$$\frac{\dot{A}}{A} = g_{a0}e^{at} \tag{5.3}$$

$$\frac{\dot{L}}{L} = g_{l0}e^{nt} \tag{5.4}$$

g_{a0} and g_{l0} are the initial growth rate of productivity level and initial growth rate of labor force, respectively. a and n are the growth rate of productivity level growth rate and the growth rate of labor force growth rate. In addition, SO_2 emission intensity and carbon intensity are dynamic, τ_{j0}^{S} and τ_{j0}^{C} denote SO_2 emission intensity and carbon intensity in 2007, respectively. λ_1 and λ_2 represent the growth rate of SO_2 emission intensity and the growth rate of carbon intensity, respectively.

$$\tau_j^S = \tau_{j0}^S e^{\lambda_1 t} \tag{5.5}$$

$$\tau_j^C = \tau_{j0}^C e^{\lambda_2 t} \tag{5.6}$$

5.2.2 *Data*

Data used for SAM table are from 2007 input-output table of China, 2007 China's statement of cash flows as well as 2008 China statistical yearbook. The SAM table in adopted this chapter, the output elasticity of capital and labor output elasticity data are from Liu (2013). SO_2 data are from China's environment statistical yearbook, the output of each department is from China's statistical yearbook. What needs to be noted here is that these departments from the two yearbooks are not correspondent to the 42 departments from input-output table. This article only consider SO_2 emissions from industrial sector (except construction), therefore based on the departments of input-output table, departments from the statistical yearbook are merged or split to obtain SO_2 emissions data and output of the resulting industrial sector except construction sector, the SO_2 emission intensity of each department is calculated. Carbon emissions and carbon intensity data of each department are from Liu (2013).

5.3 Results Analysis

5.3.1 Baseline Scenario

In this chapter, based on dynamic CGE model, China's future economic growth, SO_2 emission and carbon emission trend are forecasted, the results are taken as a benchmark scenario (BAU) to be compared with the results when levying environmental tax. In BAU scenario the GDP growth trend and the trend of SO_2 emissions are shown in Fig. 5.1.

From Fig. 5.1 it can be seen that China's GDP maintains growing under BAU scenario, rising from \$7.47 trillion in 2015 to \$69.95 trillion in 2100. SO_2 emissions showed a declining trend, which is due to technological progress and continuous decline of sulfur dioxide emission intensity. SO_2 emissions will fall from 13.57 million tons in 2015 to 3.42 million tons in 2050, in 2100 it will drop to 0.082 million tons.

From Fig. 5.2, it can be seen that carbon emissions shows obvious characteristics of EKC curve under BAU scenario, carbon emissions peak value appeared in 2034, the peak value is 3832.09 MtC, then carbon emissions continue to reduce, it will reduced to 1169.13 MtC by 2100. Zhu et al. (2009) calculated that China's future carbon emissions peak value under the optimal steady growth path is 3835.7

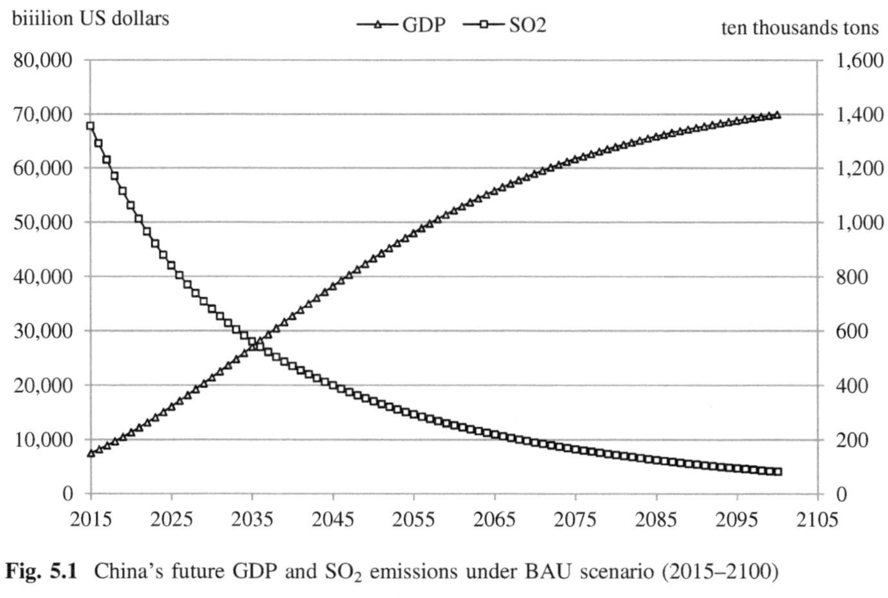

Fig. 5.1 China's future GDP and SO_2 emissions under BAU scenario (2015–2100)

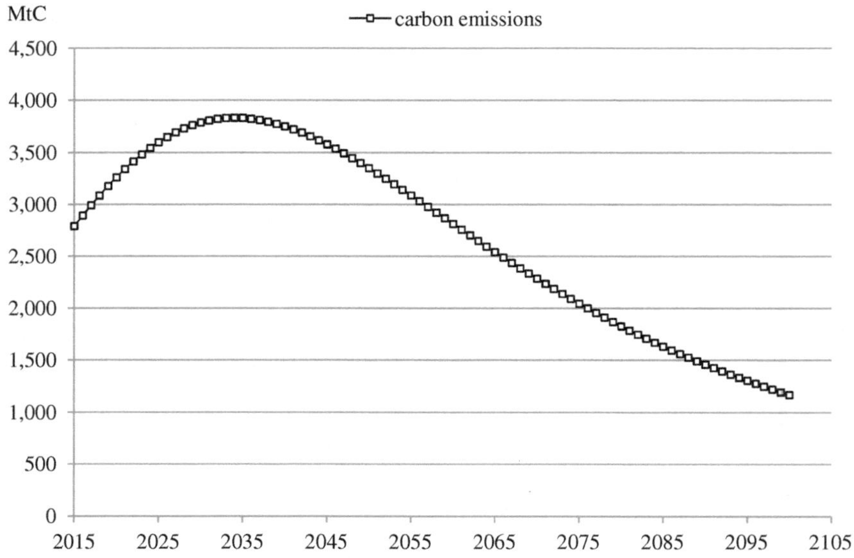

Fig. 5.2 The carbon emissions trend under BAU scenario (2015–2100)

MtC, the carbon emission peak value appeared in 2040 (Zhu et al. 2009). Wang et al. (2010) calculated China's net carbon emissions after taking account of emissions from the cement production and forest carbon sink, the peak value of net carbon emissions appeared in 2031, carbon emissions peak value is 2637 MtC (Wang et al. 2010), which is smaller than the results from CGE simulation. This is because that the CGE model and the optimal growth model is different, besides CGE simulation didn't consider the absorption effect of forest.

5.3.2 *Sulfur Tax Scenario*

Since 2005 China began to levy fees on sulfur dioxide in accordance with the standard of air pollutant. Sewage charge is 0.6 Yuan each equivalent weight of pollution, the equivalent weight of SO_2 is 0.95 kg, so sewage charge is 631.58 Yuan per ton of SO_2. Compared with the previous fee standard of 0.2 Yuan, it has improved, but still not achieving the effect of reducing SO_2 emission. Sweden started the sulfur tax in 1991 to reach the target of reducing 80% of SO_2 emissions based on 1980 year, the tax is levied according to the sulfur content of coal, per kilogram of sulfur are levied 30 Swedish kronor, which is equivalent to 14150 Yuan per ton SO_2. In this chapter, in reference to the design of Swedish sulfur tax, sulfur tax rate is designed as 14000 Yuan per ton of SO_2, sulfur tax scenario levied on the industrial sectors except construction sector were simulated.

From Fig. 5.3 It can be seen that GDP maintains a growing trend after levying sulfur tax, but compared to the baseline scenario total GDP decreased. In the sulfur tax scenario, by 2100 the total GDP would be 60.10 trillion US dollars, decreased by 9.85 trillion US dollars compared with the baseline scenario. Cumulative loss of specific departments' GDP from 2015 to 2100 are shown in Fig. 5.4.

From Fig. 5.4, it can be seen that after levying sulfur tax, the cumulative GDP loss of agriculture forestry animal husbandry and fishery is the biggest, followed by the third industrial departments, the cumulative GDP of financial services, public administration and social organizations, education, transportation and warehousing industry and real estate industry have larger loss. The second industry disaffected lightly except food manufacturing and tobacco processing industry, chemical

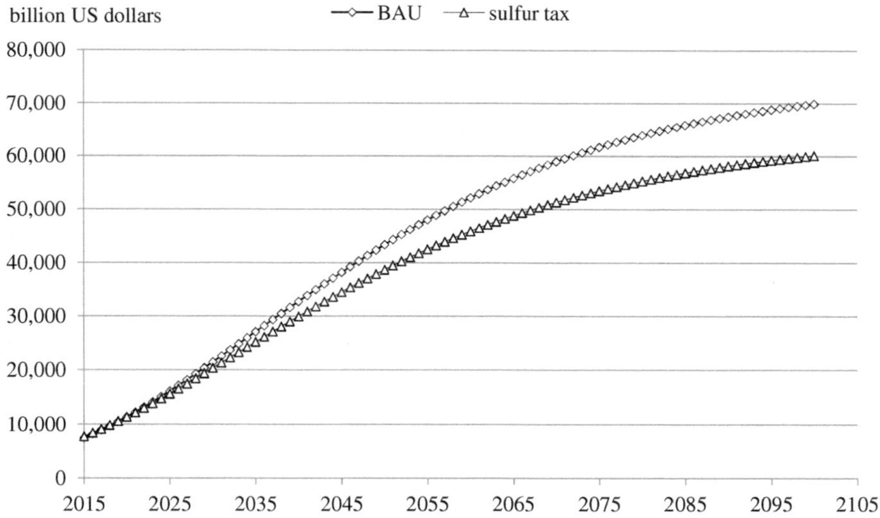

Fig. 5.3 GDP comparison between BAU scenario and sulfur scenario

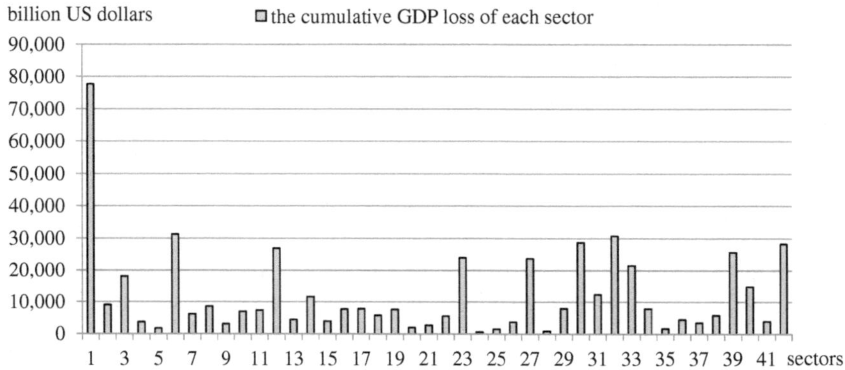

Fig. 5.4 The cumulative GDP loss of each sector in sulfur tax scenario

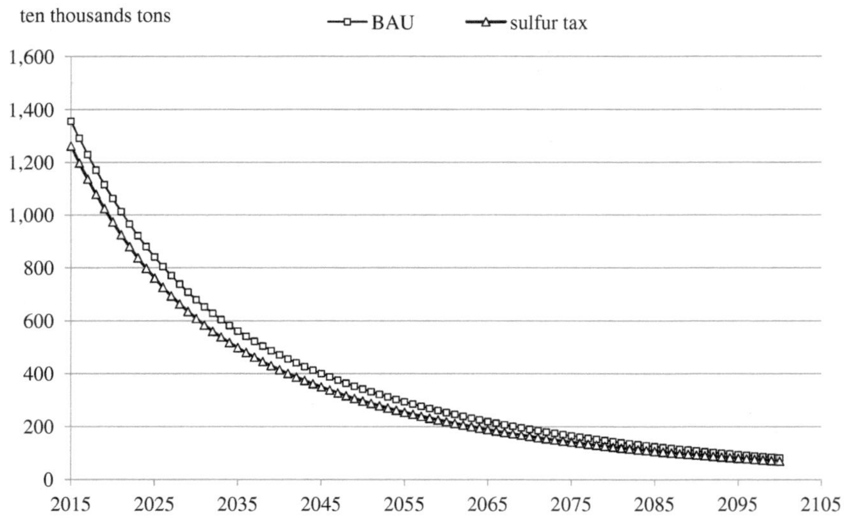

Fig. 5.5 SO_2 emissions comparison between BAU scenario and sulfur tax scenario

industry, electric power, heat production and supply industry. From the simulation results it can be seen that levying sulfur tax has larger economic impact on the primary industry and the third industry which don't emit SO_2 emission, some key SO_2 emissions emitter such as paper printing and stationery and sports goods manufacturing, non-metallic mineral products industry, metal smelting and rolling processing industry have smaller accumulated GDP losses.

From Fig. 5.5 it can be seen that after levying sulfur tax, SO_2 emission is reduced compared to the baseline scenario, emission reduction effect is remarkable especially at earlier stage. SO_2 emissions will reduce 0.94 million tons in 2015, emissions decreased gradually, in 2100, SO_2 emission will reduce 0.12 million tons.

From Fig. 5.6 it can be seen that reduction effect of sulfur tax on carbon emission is also very obvious, carbon emissions have decreased significantly compared with the baseline scenario. The peak value of carbon emissions in sulfur tax scenario is in 2032, two years earlier than the baseline scenario. The peak value of carbon emissions is 3328.39 MtC, which reduced by 503.70 MtC than the peak value in baseline scenario.

5.3.3 Carbon Tax Scenario

Considering foreign mature carbon tax mechanism and the carbon tax simulation of domestic scholars of our country, this chapter simulates the impact of carbon tax on the economy and emission reduction effect, the tax rate is set to 50 Yuan per ton of carbon.

From Fig. 5.7 it can be seen that after levying carbon tax GDP declined compared with the baseline scenario, in 2100 GDP is 64.54 trillion US dollars, which

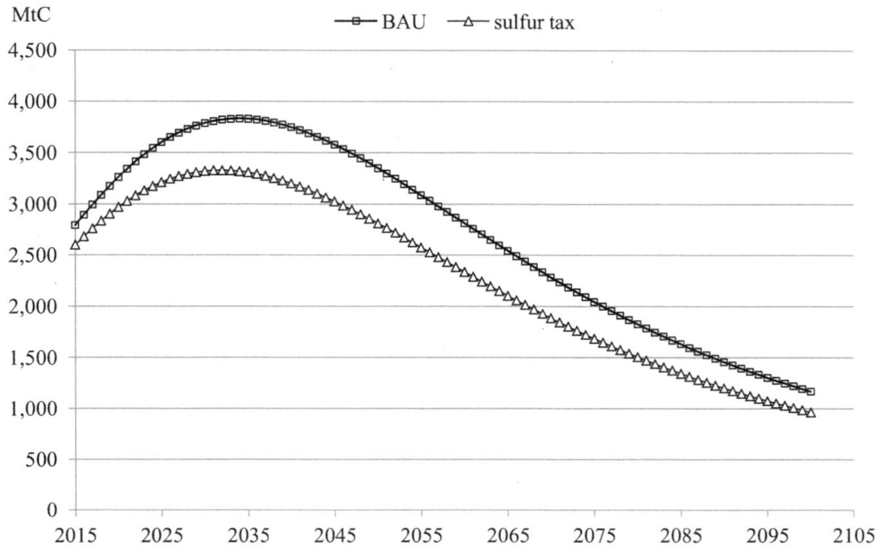

Fig. 5.6 Carbon emissions comparison between BAU scenario and sulfur tax scenario

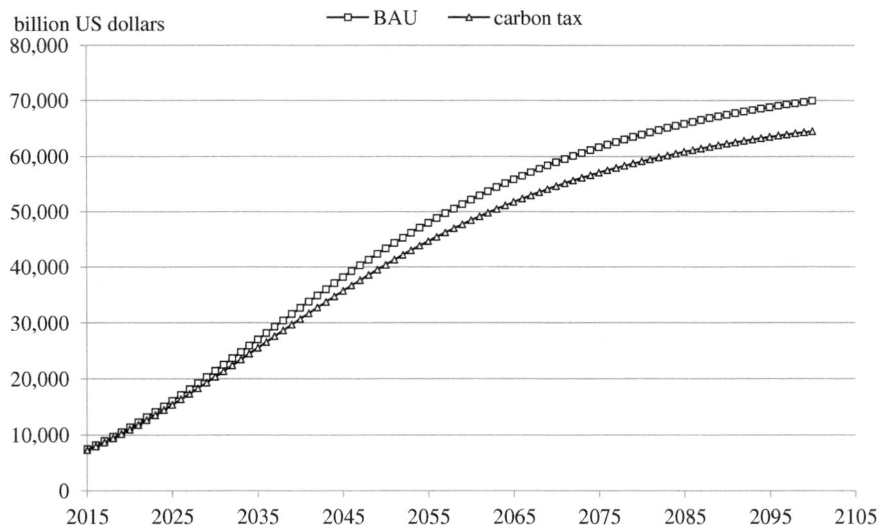

Fig. 5.7 GDP comparison between BAU scenario and carbon tax scenario

declined by 5.42 trillion US dollars compared to the baseline scenario. GDP losses of specific departments are shown in Table 5.1.

The cumulative GDP losses of all departments, GDP losses in 2030, 2050 and 2100 after levying carbon tax can be seen from Table 5.1. Simulation results show

Table 5.1 GDP loss of each sector in carbon tax scenario (%)

No.	Sectors	Cumulative	2030	2050	2100
1	Agriculture, forestry, animal husbandry and fishery	−7.39	−5.37	−7.06	−8.04
2	Coal mining and washing industry	−8.37	−6.18	−8.05	−9.15
3	Petroleum and natural gas exploitation	−8.14	−6.16	−7.83	−8.8
4	Metal mining industry	−7.22	−5.23	−6.91	−8
5	Nonmetallic ore and other mining industry	−6.86	−4.77	−6.56	−7.71
6	Food manufacturing and tobacco processing industry	−7.37	−5.38	−7.04	−8
7	Textile industry	−7.1	−4.84	−6.78	−7.91
8	Textile, leather and feather products industry	−7.51	−5.39	−7.19	−8.22
9	Wood processing and furniture manufacturing	−6.82	−4.66	−6.51	−7.64
10	Paper printing and sports goods manufacturing	−7.06	−5.08	−6.77	−7.73
11	Petroleum processing, coking and nuclear fuel processing industry	−7.91	−5.8	−7.6	−8.65
12	Chemical industry	−7.55	−5.5	−7.24	−8.26
13	Nonmetallic mineral products industry	−6.25	−4.24	−5.96	−7.2
14	Metal smelting and pressing industry	−6.76	−4.66	−6.47	−7.65
15	Fabricated metal products	−6.96	−4.8	−6.66	−7.82
16	General, special equipment manufacturing	−6.45	−4.39	−6.16	−7.36
17	Transportation equipment manufacturing	−7.04	−4.87	−6.73	−7.88
18	Electrical machinery and equipment manufacturing	−7.3	−5.14	−7	−8.14
19	Communications equipment, computers and other electronic equipment manufacturing	−6.72	−4.57	−6.43	−7.55
20	Instrumentation and cultural office supplies machinery manufacturing	−7.35	−5.3	−7.05	−8.08
21	Handicrafts and other manufacturing	−7.64	−5.45	−7.32	−8.41
22	Waste	−7.13	−5.11	−6.83	−7.89
23	Electricity, heat production and supply industry	−9.49	−7.26	−9.18	−10.26
24	Gas production and supply	−8.27	−6.11	−7.94	−8.99
25	Water production and supply industry	−7.9	−5.9	−7.59	−8.56
26	Construction	−4.02	−3.01	−3.87	−4.76
27	Transportation and warehousing	−7.38	−5.29	−7.07	−8.11
28	Postal services	−7.28	−5.32	−6.98	−7.93
29	Information transmission, computer services and software industry	−4.97	−3.73	−4.86	−5.35
30	Wholesale and retail trade	−6.49	−4.59	−6.21	−7.14
31	Accommodation and catering services	−6.35	−4.67	−6.11	−6.89
32	Financial industry	−6.99	−4.95	−6.68	−7.67
33	Real estate	−5.05	−3.45	−4.81	−5.57
34	Leasing and business services	−6.97	−5.01	−6.67	−7.62

(continued)

Table 5.1 (continued)

No.	Sectors	Cumulative	2030	2050	2100
35	Research and experimental development industry	−7.09	−5.32	−6.82	−7.68
36	Integrated technical services	−7.26	−5.19	−6.96	−8.01
37	Water conservancy, environment and public facilities management industry	−7.31	−5.29	−6.97	−7.96
38	Resident services and other services	−5.11	−4.06	−5.03	−5.45
39	Education	−7.37	−5.45	−7.05	−7.98
40	Health, social security and social welfare	−7.94	−6.01	−7.63	−8.56
41	Culture, sports and entertainment	−6.81	−4.98	−6.52	−7.41
42	Public management and social organization	−7.25	−5.36	−6.95	−7.85

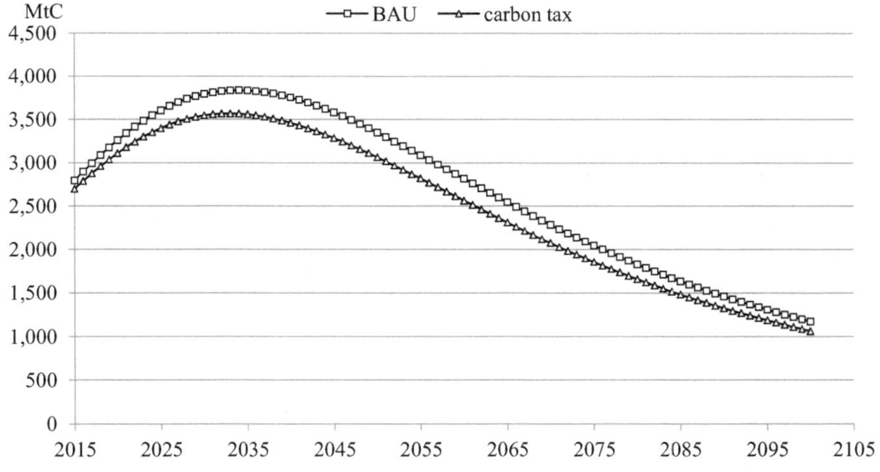

Fig. 5.8 Carbon emissions comparison between BAU scenario and carbon tax scenario

that each sector's output are affected negatively after levying carbon tax, among them, the negative influence on electricity, heat production and supply industry is the largest, whose GDP loss in 2100 reached 10.26%. Followed by coal mining and coal washing industry, whose GDP loss in 2100 is 9.15%. The GDP losses of petroleum and natural gas exploitation, oil processing, coking and nuclear fuel processing, production and supply of gas and water production and supply industry are also large, which were 8.80, 8.65, 8.99, 8.56%, respectively.

From Fig. 5.8 it can be seen that carbon emissions decreased than the baseline scenario after levying carbon tax. In carbon tax scenario, carbon emissions peak is in 2033, the peak value is 3564.65 MtC, which reduced by 267.45 MtC than the emissions peak value in the baseline scenario.

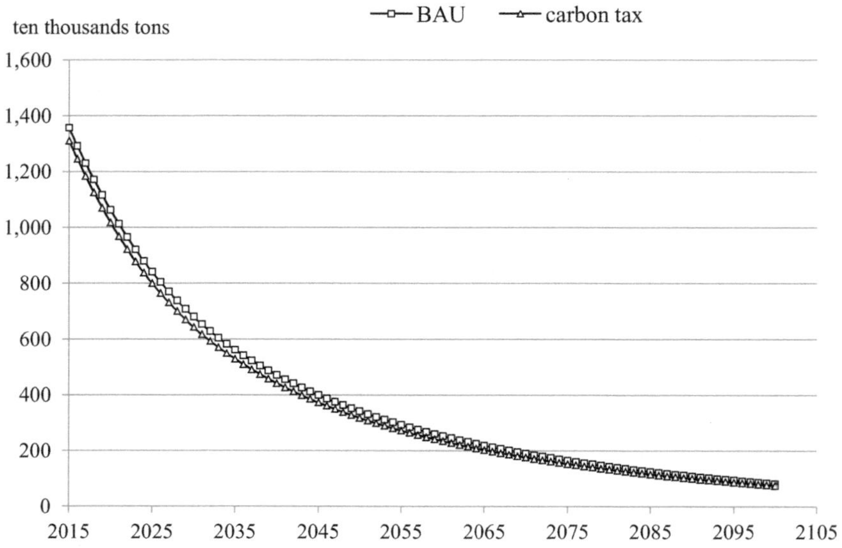

Fig. 5.9 SO_2 emissions comparison between BAU scenario and carbon tax scenario

From Fig. 5.9 it can be seen that carbon tax can reduce SO_2 emissions, SO_2 emissions will reduce by 0.47 million tons in 2015, then decreased gradually, in 2050 it will reduce by 0.24 million tons, it will reduce by 0.06 million tons in 2100.

5.3.4 Sulfur Tax and Carbon Tax Scenario

Levying sulfur tax and carbon tax at the same time were simulated in order to get economic effects of sulfur tax and carbon tax and reduction effect.

From Fig. 5.10 it can be seen that GDP decreased significantly compared with the baseline scenario after levying sulfur tax and carbon tax at the same time, the negative impact is gradually increasing, GDP is 55.12 trillion US dollars in 2100, which reduced by 14.83 trillion US dollars, dropped more than 21.20% compared with the baseline scenario.

From Fig. 5.11 it can be seen that levying both sulfur tax and carbon tax can significantly reduce SO_2 compared with the baseline scenario, especially in the early stage.SO_2 emissions will be reduced by 1.30 million tons in 2020, emission reduction rate reached to 12.26%. In 2050 emission reductions amount reached 0.65 million tons, reduction rate is 19.16%. SO_2 emissions will be reduced to 0.18 million tons in 2100, reduction rate is 21.85%.

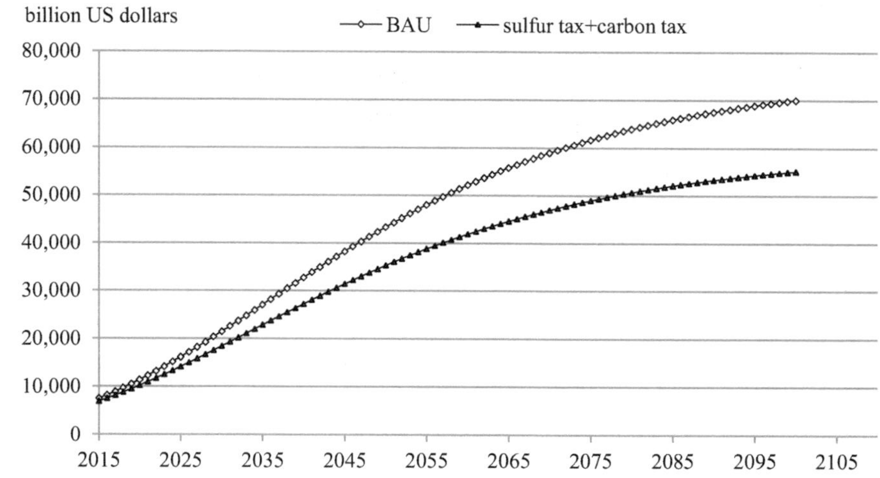

Fig. 5.10 GDP comparison between BAU scenario and sulfur tax and carbon tax scenario

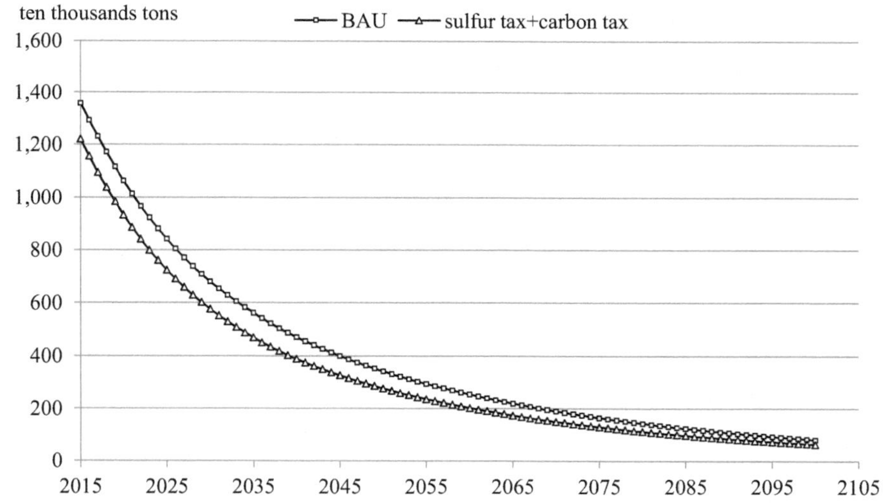

Fig. 5.11 SO_2 comparison between BAU scenario and sulfur tax and carbon tax scenario

From Fig. 5.12 it can be seen that carbon emissions are significantly lower than the baseline scenario after levying carbon tax and sulfur tax at the same time. Carbon emission speak appeared in 2031, the peak value is 3111.11 MtC, which decreased 720.99 MtC compared with the baseline scenario, the peak year is three years earlier than the baseline scenario.

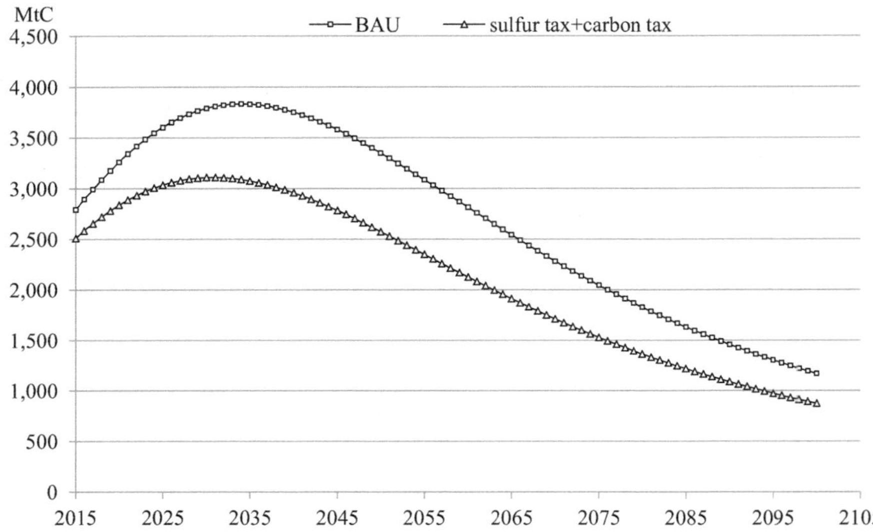

Fig. 5.12 Carbon emissions comparison between BAU scenario and sulfur tax and carbon tax scenario

5.4 Conclusions

In this chapter, based on a dynamic computable general equilibrium model the economic development trend and reduction effects of levying sulfur tax, carbon tax, both of sulfur tax and carbon tax at the same time were simulated, respectively. The main conclusions are as follows:

(1) In the baseline scenario, China's economy will maintain a rising trend, the total GDP will reach 69.95 trillion US dollars by 2100, SO_2 will continue to reduce emissions due to technological progress and sulfur dioxide emission intensity decrease. Carbon emissions present the characteristics of EKC, the peak value appeared in 2034, carbon emission peak value is 3832 MtC. (2) In the sulfur tax scenario, GDP has declined compared with the baseline scenario, the total GDP is 60.10 trillion US dollars in 2100. Levying sulfur tax not only reduces the SO_2 emissions, but also decreases carbon emissions significantly compared with the baseline scenario, carbon emissions peak value appeared in 2032. (3) In the carbon tax scenario, GDP is 64.54 trillion US dollars in 2100, which declined by 5.42 trillion US dollars compared to the baseline scenario. Carbon emissions peak appears in 2033, the peak value is 3564.65 MtC, which reduced by 267.45 MtC than the emissions peak value in baseline scenario. (4) In the sulfur tax and carbon tax scenario, GDP is 55.12 trillion US dollars in 2100, which

reduced by 14.83 trillion US dollars than the baseline scenario, SO_2 emission and carbon emissions were reduced, carbon emissions peak appeared in 2031.

Some scholars proposed that the environmental tax can bring double dividend (Gao and Li 2009). Based on the conclusions in this chapter, sulfur tax, carbon tax and levying these two environmental taxes at the same time did not lead to double dividend. Thus we give the following policy suggestions: (1) the tax rate design of the environmental tax should be in line with China's actual conditions. In order to reduce emissions and protect the environment, economic development should be considered not to produce great economic loss. (2) The environment taxes should be reasonable, to avoid negative effects of double or even multiple taxes on the economy.

References

Cao J (2009) Low carbon development road: the design and CGE analysis model of China's carbon tax policy. Finan Res 12:19–29

Gao PF, Chen WY (2002) Carbon tax and carbon emission. J Tsinghua Univ (Sci &Tech) (10):1335–1338

Gao Y, Li ST (2009) Analysis on the influence of energy tax on society, economy, energy and environment. China Popul Resour Environ 12(2):30–35

He JH, Shen KT, Xu SL (2002) Carbon tax and carbon dioxide emissions of CGE model. Technol Econ Res Quant Econ 10:39–47

Li C (2014) Policy simulation of carbon tax based on CGE model. J Ind Technol Econ 01:146–153

Liu CX (2013) The construction of a new style of IAM and study on the global corporations for the mitigation of the carbon dioxide. Doctoral dissertation, Institute of Policy and Management, Chinese Academy of Sciences

Liu CX, Wang YF, Hao ZP et al (2013) Study of VOCs emission prediction and control based on Dynamic CGE. Environ Sci 34(12):4787–4791

Ma SG (2008) Effects of sulfur tax on Chinese sulfur dioxide emission and energy consumption. China Ind Econ 2:20–30

Tu ZG (2009) The shadow price of industrial SO_2 emission: a new analytic framework. China Econ Q 1:259–282

Wang C, Chen JN, Zou J (2005) Impact assessment of CO_2 mitigation on China economy based on a CGE model. J Tsinghua Univ (Sci &Tech) (12):1621–1624

Wang JN, Yan G, Jiang KJ et al (2009) The study on China's carbon tax policy to mitigate climate change. China Environ Sci 01:101–105

Wang Z, Xue JB, Zhu YB et al (2010) CGE technology of economic development policy simulation analysis. Science Press, Beijing

Wang Z, Zhu YB, Liu CX et al (2010b) Integrated projection of carbon emission for China under the optimal economic growth path. Acta Geographica Sinica 65(12):1559–1568

Wu YJ, Xuan XW (2002) Environmental taxes and application of economic theory analysis of China. Economic Science Press, Beijing

Yuan P, Cheng S (2011) Estimating shadow pricing of industrial pollutions in China. Stat Res 28 (9):66–73

Zhu YB, Wang Z, Pang L et al (2009) Simulation on China's economy and prediction on energy consumption and carbon emission under optimal growth path. Acta Geographica Sinica 64 (8):935–944

Zhu YB, Liu X, Wang Z (2010) Abatement effect of carbon tax and its impacts on economy in China. China Soft Sci 04:1–9

Chapter 6
CIECIA

6.1 Introduction

Currently, our world is in a serious situation regarding climate change, making it urgent to implement appropriate solutions. Numerous global carbon abatement schemes have been proposed including the Stern Scheme, Nordhaus Scheme, UNDP Scheme and some important strategies proposed by Chinese scholars e.g. the principle of equality on global per capita emissions, the principle of equality on accumulated per capita emissions and the Global Economic Growth Scheme (He et al. 2004; Chen et al. 2005; Ding et al. 2009; Wang et al. 2012; UNDP 2007). However, there are still divergences in reduction responsibilities and obligations among countries. The primary reason for these divergences is that the global cooperating carbon abatement is closely related to the economic loss division and the distribution of economic benefits of countries. Thus, it is important to build a global economic-climatic integrated model to accurately assess the economic impact of the climate mitigation scheme on different countries, searching for reasonable carbon abatement schemes that consider fully the economic benefits to abatement participating countries, and providing scientific bases for the international climate negotiations.

Integrated Assessment Model (IAM) has become mainstream in global economic-climatic research studies. IAMs, which combine interdisciplinary theories and methods, and have comprehensive and integrated system structures and powerful policy assessing abilities, synthetically consider human economic activities and climate change, and are very applicable to simulation and analysis of global climate mitigation policies (Weyant et al. 1996; Whitman et al. 1997; Schneider 1997; Kelly and Kolstad 1999; Tol 2001; Ackerman et al. 2009). However, the existing IAMs still have some shortcomings and are facing challenges.

First, in many existing IAMs, including some famous ones such as DICE (Nordhaus 1992) and RICE (Nordhuas and Yang 1996), the economies of countries

Z. Wang et al., *Integrated Assessment Models of Climate Change Economics*, DOI 10.1007/978-981-10-3945-4_6

are respectively independent, which makes those IAMs unable to reflect the impact of one country's economic change on the global economy. Among them, DICE considered the world economy as a whole; RICE divided the global economy into country level; however, the international economic relationships are still lacking; IMAGE (Rotmans 1990) divided the economy into sectoral level; however, there is still a lack of inter-sectoral economic relationships. Thus, the mitigation schemes developed by those IAMs do not conform to reality because they ignore international economic relationships. Actually, the schemes generated by those IAMs always contain risks leading to global economic crises because of the possibility of breaking the global economic general equilibrium.

Second, the process of technological progress is essential for future development. In fact, technological innovation has been considered the most effective solution to environmental problems. Thus, ignoring technological progress does not conform to reality. Technological progress would reduce energy consumption and the emissions of pollutants and reconstruct the industrial structures of countries. Unfortunately, some existing IAMs, especially policy optimization IAMs, e.g., MERGE (Manne et al. 1995) and WITCH (Bosetti et al. 2007), trend to pay attention to energy technologies, and always ignore changes in industrial structure influenced by technological progress, which is also an important basis of carbon emission intensity reduction. In addition, policy optimization IAMs are always relatively simple for their computational expenditures. Thus, their economic divisions are always limited at the country level and cannot reflect the economic relationships among sectors. MRICES (Wang et al. 2012), which was developed based on RICE, adopts an international GDP spillover model to describe economic relationships among countries. However, its economic division is still at the country level, making it impossible to describe the technological progress of the carbon-involving industries exactly. In addition, because different sectors always have different emission behaviors, the generalized macro models cannot reflect relative environmental problems appropriately.

Third, the technological progress mechanisms in some IAMs always adopt methods such as the empirical estimation of progress speeds and technology parameter estimation. These methods cannot reflect the uncertainty of technology progress, causing those macroeconomic models to degenerate into technical economic models that are utilized for short-time problem analysis and decreasing the reliability of the climatic-economic assessments. Because of parameter sensitivities, the outcomes of those models are unstable.

Fourth, some technical economic models consider neither the global economic equilibrium, nor the industrial structure change, whereas CGE models, e.g. GREEN, FUND, EPPA, and ANEMI_2, focus on the global economic equilibrium, but simplify the economic impact of climate change too much (Bosetti et al. 2007; OECD 1997; Tol 1997; Paltsev et al. 2005; Akhtar et al. 2013). Thus, those models are not very realistic.

To overcome the shortcomings of the models mentioned above, the global GHG cooperating abatements and their economic impacts should be studied in the background of industrial change and global economic general equilibrium. In this study, a new climate change economic IAM, named the Capital, Industrial Evolution and Climate change Integrated Assessment model (CIECIA) has been built. The economic core of CIECIA is a multi-country-sector general equilibrium model that is modified and extended based on Jin (2012). Two international capital flow modes are adopted in CIECIA for reflecting the economic interactions among sectors and countries under the constraints of equilibrium. Meanwhile, according to the mode of evolutionary economics (Nelson and Winter 1982), the theories of process technological progress and knowledge capital are introduced into CIECIA to realize the endogenous mechanism of technological progress (Lorentz and Savona 2008; Buonanno et al. 2003). Based on CIECIA, carbon abatement scenarios are designed to assess and analyze the global cooperating carbon abatement schemes in this study.

6.2 Model and Data Sources

For solving the problems proposed here, modeling innovations are necessary. First, the global economic equilibrium should be included in the model system to reflect economic globalization and to ensure that the global economy will not be at risk because of the implementation of the abatement policies of several countries. Second, to reflect the real world, the endogenous sectoral technology progress mechanism should be realized in the model, ensuring the policy analysis of the sectoral innovation and climate policy. The new IAM that is built in this research—CIECIA—meets those requirements. CIECIA divides the entire world into several countries, and in every country, the economic system is composed of several sectors. Every sector in the countries has several economic behaviors, e.g., producing, trading, consuming, investing, and realizing technology progress, etc., and the global economic equilibrium is achieved based on the demand-supply equilibrium among sectors of countries. It is assumed that the capital and commodity flow among sectors of countries freely, without investment or trade barriers. Sectors in one country have a unique wage rate and the same labor technology level in one step. The discount rates of countries are uniform. In this study, henceforth, j denotes countries, i denotes sectors and g denotes global level.

6.2.1 *Economic Module*

In this study, a two-layer-nested function formation comprises the Leontief function and the Cobb-Douglas function is employed to represent the relationships among labor, capital stock, value added and gross output. The sectoral gross output is composed of value added and intermediate inputs, and the value added is formed by labor, capital stock and knowledge capital.

$$X_{i,t}^{j} = \min\left\{\frac{M_{1,i,t}^{j}}{a_{1,j,i,t}}, \ldots, \frac{M_{k,i,t}^{j}}{a_{k,j,i,t}}, \ldots, \frac{M_{I,i,t}^{j}}{a_{I,j,i,t}}, X_{i,j,t}^{*}\right\}, \quad k = 1, \ldots, I \tag{6.1}$$

$$X_{i,j,t}^{*} = \Omega_{i,j,t}\left(K_{i,t}^{j}\right)^{\alpha_i}\left(A_t^j L_{i,t}^{j}\right)^{1-\alpha_i}\left(Z_t^j\right)^{\beta_{j,t}^{Z}} \tag{6.2}$$

$$Y_{i,t}^{j} = \sum_{k}^{I} M_{k,i,t}^{j} p_{k,t} + X_{i,t}^{j} \tag{6.3}$$

where $X_{i,t}^{j}$ represents the value added of sector i of country j in step t; $X_{i,j,t}^{*}$ represents the initial value added formed by labor and capital; $M_{i,t}^{j,k}$ is the intermediate input k in the production process of sector i of country j; $a_{k,j,i,t}$ is the coefficient of intermediate input of sector i in country j; $K_{i,t}^{j}$ is the capital stock of sector i in country j in step t; $L_{i,t}^{j}$ denotes the labor; A_t^j is the labor technology level of country j in step t; α_i is the output elasticity of capital stock of sector i; Z_t^j denotes the knowledge stock of country j in step t; $\beta_{j,t}^{Z}$ denotes the output elasticity of knowledge capital; $p_{k,t}$ is the price of good k. Because of free trade among countries, there is one international price associated with each good i in each step; $\Omega_{i,j,t}$ denotes the influence of climate change on the economy, which will be described in Sect. 2.2. The purpose of introducing the Leontief function is to implement the abatement policies into sectors.

The associated price index of goods is composite by the good's price $p_{i,t}$ and the global output equilibrium share parameter γ_i. According to Jin (2012), the associated price index is the same in all countries and is normalized to 1.

$$\bar{p}_t = \prod_{i}^{I} \left(p_{i,t}\right)^{\gamma_i} = 1 \tag{6.4}$$

The wage rate equals the marginal output of labor. Because of the full mobility of workers in one country, there is a unique wage rate across sectors in this country in one step.

$$w_{i,t}^{j} = \frac{\partial X_{i,t}^{j}}{\partial L_{i,t}^{j}} p_{i,t} = (1 - \alpha_i)\frac{X_{i,t}^{j}}{L_{i,t}^{j}} p_{i,t} \tag{6.5}$$

According to Abel (2003) and Jin (2012), the cumulative process of capital stock is represented in the form of the Cobb-Douglas production function in this study, whereas the knowledge capital is accumulated with the Perpetual Inventory Method.

$$K_{i,t+1}^{j} = a(I_{K\,i,t}^{j})^{\phi}(K_{i,t}^{j})^{1-\phi} \tag{6.6}$$

$$Z_{t+1}^{j} = (1-\delta_Z)Z_t^j + I_{Zt}^{j} \tag{6.7}$$

$$\sum_j^J I_{Zt}^j + \sum_j^J \sum_i^I I_{K\,i,t}^j = I_t^g \tag{6.8}$$

where $I_{K\,i,t}^{j}$ denotes the investment to capital stock of sector i in country j; I_{Zt}^{j} denotes the knowledge capital investment of country j; I_t^g is the global investment in step t. In this study, the international investing behavior is a weighting form of two different investing mode following Gu and Wang (2014).

The good markets clear when the global output of each product equals its global consumption, which comprises intermediate consumption, final consumption and the part for capital goods commodities (CGDS).

$$Y_{i,t}^{g} = \sum_j^J Y_{i,t}^j = \sum_j^J c_{i,t}^j + \sum_j^J \sum_i^I x_{k,i,t}^j + \sum_j^J \sum_i^I \sum_k^I M_{k,i,t}^j p_{i,t} \tag{6.9}$$

The global value added comprises the labor value added and the capital value added, which can be divided into consumption and investment on the demand side. Thus, the global value added clearing is given by

$$X_{i,t}^{g} = \sum_j^J X_{i,t}^j = \sum_j^J c_{i,t}^j + \sum_j^J \sum_k^I x_{k,i,t}^j \tag{6.10}$$

According to Jin (2012), the price equilibrium function can be given by

$$p_{i,t} = \frac{\gamma_i}{X_{i,t}^g} X_t^g \tag{6.11}$$

From Eq. 6.4 to Eq. 6.11, a global economic equilibrium system is built, which ensures that the implementation of policies is restricted to not destroying economic equilibrium in the analysis of global carbon abatement policies and avoiding causing economic risks or stimulating crises.

According to Wang et al. (2012), the Keynes-Ramsey utility function is adopted in this study to reflect the changes of economic powers of countries, and is taken as the basis of the analysis of abatement schemes' economic impacts.

$$UA^j(T) = \sum_{t=1}^{T} (\beta+1)^{-t} Pop_t^j \frac{\left(C_t^j / Pop_t^j\right)^{1-\rho}}{1-\rho} \tag{6.12}$$

where $UA^j(T)$ denotes the accumulated Ramsey utility of country j until step T; C_t^j denotes the consumption of country j in step t; Pop_t^j denotes the population of country j; β is the discount rate; ρ is the time preference of consumers. According to Buchner and Carraro (2005) and Wang et al. (2012), the value of β is 0.015, and according to Wang et al. (2006), the value of ρ is 0.198, which is gained by computing the depreciation rates of Currencies and the changes in exchange rates.

6.2.2 Climate Module

All of the energy products are provided by the energy sector of countries. The energy consumption of sectors in countries can be given by

$$E_{i,t}^{j} = \tau_{i,j}^{E} M_{E,i,t}^{j} \tag{6.13}$$

where $E_{i,t}^{j}$ denotes the energy consumption of sector i in country j in step t; $\tau_{i,j}^{E}$ denotes the energy consumption per energy product use, i.e. energy intensity. Thus, the carbon emissions of countries can be given by

$$QP_t^j = \sum_e \tau_{j,e}^{C} \kappa_{e,i,t} \sum_i E_{i,t}^{j} \tag{6.14}$$

where QP_t^j denotes the carbon emissions of country j in step t; $\tau_{j,e}^{C}$ denotes the carbon intensity of energy e in country j. In this study, fossil energy comprises oil, coal and gas, and their ratios are obtained by fitting the energy consumption data from EIA.

In addition, the carbon emissions from land use change must also be considered in the accounting of global carbon emissions.

$$Q_t = \sum_j^J QP_t^j + QL_{t0}(1 - \gamma_{le})^t \tag{6.15}$$

where Q_t is the global carbon emissions in step t; QL_{t0} denotes the initial value of global carbon emissions from land use change; γ_{le} denotes the yearly decay rates of global carbon emissions from land use change.

According to Wu et al. (2014), the Svirezhev carbon-cycle model (S-N model) which contains a terrestrial ecosystem turns out to be superior to the solo-reservoir and three-reservoir models of DICE/RICE in terms of a much more detailed model mechanism and more accurate modeling performance. Thus, the Svirezhev carbon-cycle model is adopted in CIECIA. The S-N model divided the carbon-cycle process into the terrestrial carbon cycle, ocean carbon cycle and atmospheric carbon cycle, among them, the terrestrial carbon is shared between two compartments: biota (vegetation) and pedosphere (soils). The changes in atmospheric carbon content can be given by

$$\Psi_t = \Psi_{t-1} + Q_t - \Delta V_t - \Delta So_t - \Delta O_t \tag{6.16}$$

where Ψ_t denotes the atmospheric carbon content in step t, ΔV_t denotes the net change of the amount of carbon in vegetation in step t; ΔSo_t denotes the net change of the amount of carbon in soils; ΔO_t denotes the net change of the carbon content in ocean.

According to Nordhuas and Yang (1996) and Pizer (1999), the climate damage function which reflects the economic impact of climate change is given by

$$\Omega^i_{j,t} = \frac{1 - b_{1,j}\mu^j_t}{1 + D_{0,j}\left(\frac{T_t}{3}\right)^2} \tag{6.17}$$

where $b_{1,j}$ denotes the production damage coefficient of country j; $D_{0,j}$ is the GDP loss from 3 °C of global warming; T_t is the global mean surface temperature in step t, μ^j_t denotes the carbon reduction rate of country j; $1 - b_{1,j}\mu^j_{i,t}$ denotes the economic loss caused by the carbon abatement.

6.2.3 Technological Progress

Technology progress is an economic fact that needs to be considered in the research of the long-term development problems. One of the most common results of technological progress is the reduction of intermediate demand in the production process.[1] In this study, according to Lorentz and Savona (2008), a mechanism of stochastic logarithmic shock is adopted for the changes of the coefficient of intermediate demand $a_{k,i,j,t}$.

$$\ln(a'_{k,j,i,t}) = \ln(a^*_{k,j,i,t,n-1}) - \varepsilon_{j,k,i,t,n} \tag{6.18}$$

$$\varepsilon_{j,k,i,t,n} \in N(0;\rho^j_t) \tag{6.19}$$

where $\varepsilon_{j,k,i,t,n}$ is a stochastic number following normal distribution $N(0;\rho^j_t)$. A new set of coefficients $\left(a'_{1,j,i,t},\ldots,a'_{I,j,i,t}\right)$ is generated after one iteration of stochastic shock. Then, the new set is accepted to replace the original one if the potential unitary cost is lower than the original unitary cost $\sum^J_k a^*_{k,j,i,t,n-1}p_{i,t}$.

$$\left(a^*_{1,j,i,t,n},\ldots,a^*_{I,j,i,t,n}\right) = \begin{cases} \left(a'_{1,j,i,t},\ldots,a'_{I,j,i,t}\right) & \text{if } \sum\limits^J_k a'_{k,j,i,t}p_{i,t} < \sum\limits^J_k a^*_{k,j,i,t,n-1}p_{i,t} \\ \left(a^*_{1,j,i,t,n-1},\ldots,a^*_{I,j,i,t,n-1}\right) & \text{Otherwise} \end{cases} \tag{6.20}$$

[1]Technological progress here specifically refers to the process technological progress. Process technological progress refers to the innovation of the production process, reflecting the reduction of unitary intermediate cost. Although this concept of process technological progress ignores revolutionary product innovation, it accords more with the history of economic development.

To overcome the uncertainty caused by the contingency in the process of technological progress, there are 1000 iterations of stochastic shock in one step. After the 1000th iteration, $\left(a^*_{1,j,i,t,n},\ldots,a^*_{I,j,i,t,n}\right)$ is introduced in the production function for the next step. The variance ρ_t^j of stochastic shock determines the pace of technological progress. In this study, it is assumed that the ratio of knowledge capital in the total capital of one country affects the rate of process technological progress.

$$\rho_t^j = a^\rho \frac{Z_t^j}{Z_t^j + K_t^j} + b^\rho + c_{j,t}^\rho \tag{6.21}$$

where ρ_t^j depicts the possibility of technological progress; a^ρ, b^ρ, $c_{j,t}^\rho$ are the parameters in the function. Among them, $c_{j,t}^\rho$ is an adjustment to the rate of process technological progress of country j. In Eq. 6.21, the possibility of technological progress increases with the ratio of knowledge capital. Thus the endogenous mechanism of technological progress is formed, for the knowledge capital is endogenous.

This mechanism of technological progress reflects the mode of the evolutionary economics of Nelson and Winter (1982). The technological progress confirmed by Eq. 6.20 is not only a slow natural process but also the profit choices of firms. In addition, it is influenced by stochastic interferences. It reflects a type of randomness in the technological process itself so that the stochastic interference $\varepsilon_{j,k\,i,t,n}$ has a normal distribution.

6.2.4 Data Sources

The primary source of the economic data of CIECIA including the initial GDPs, gross outputs, capital stocks, intermediate input coefficients and energy intensities of sectors and countries is GTAP-07 database (Narayanan and Walmsley 2008). The populations of countries, as well as their growth rates are obtained from *World Population Prospects: The 2010 Revision* (United Nations Department of Economic and Social Affairs 2011). The values of the parameters about the knowledge capital are cited from Wang et al. (2012), Zhang (2012) and Liu (2013). The values of the parameters of the global carbon-cycle model are cited from Svirezhev et al. (1999), Zhu (2012) and Wu et al. (2014). The parameters of the damage function are cited from Nordhuas and Yang (1996), Pizer (1999) and Zhu (2012). The carbon emissions and the carbon intensities of different types of energy of countries are obtained from the website of EIA.[2] The capital output elasticities of sectors are calculated based on the added values and wages obtained from

[2]EIA. http://www.eia.gov/.

Table 6.1 The capital output elasticities of sectors

Sector	The capital output elasticity
Agriculture	0.2512
Food processing	0.4514
Energy	0.5997
Metal and other minerals	0.4601
Light manufacturing	0.4104
Chemical industry	0.5166
Heavy manufacturing	0.3942
Construction	0.3515
Trade and business services	0.4914
Transport and communication	0.5217
Insurance and finance services	0.5858
Other services	0.4705

Table 6.2 Parameter of the relationship between Tech-shock and knowledge Capital (1×10^{-5})

	CHN	USA	JPN	EU	IND	RUS	ODC	HDC	MDC	LDC
$c^{\rho}_{j,t}$	2.7486	3.1119	−1.8432	2.0534	0.4239	2.4382	−0.4171	−5.5618	−0.5761	−2.3780

GTAP-07 (Table 6.1). The other main parameters values and initial values are listed in Appendix A.

In addition, in Eq. 6.20, the values of a^{ρ} and b^{ρ} are 5.2298×10^{-3} and 1.0499×10^{-4} respectively. Table 6.2 presents the values of $c^{\rho}_{j,t}$. In the regression analysis of Eq. 6.20, R^2 is 0.8740, and the t statistics of the two parameters are −2.89 and 7.75 respectively. The hypothesis of Eq. 6.20 passes the test.

For simplification, this study merges 57 sectors of GTAP07 into 12, comprising Agriculture, Food Processing, Energy, Metal and other Minerals, Light Manufacturing, Chemical industry, Heavy Manufacturing, Construction, Trade and Business Services, Transport and Communication, Insurance and Finance Services and Other Services after considering their energy consumptions and supply characteristics. CIECIA divided the world into 10 countries/regions, comprising China (CHN, for short), the United States (USA), Japan (JPN), the European Union (EU), Russia (RUS), India (IND), Other developed countries (ODC), High developing countries (HDC), Middle developing countries (MDC), and Low developing countries (LDC), as the same as MRICES-2012. One step is chosen to be one year in this study.

6.3 Calibration

Table 6.3 presents the comparison of the GDP of various countries from 2007 to 2011 between the real data obtained from the EIA and the simulation outcomes. In the regression analysis, the correlation coefficient between these two sets of data is

Table 6.3 Calibration of GDP (10^9US Dollars, in constant 2006 prices)

		CHN	USA	JPN	EU	IND	RUS	ODC	HDC	MDC	LDC
EIA Data	2007	2904.59	13206.27	4753.29	14630.98	1004.75	896.86	4696.43	5286.00	1524.96	681.44
	2008	3183.35	13161.99	4697.59	14659.69	1054.31	943.92	4772.28	5478.94	1603.04	720.78
	2009	3476.64	12757.97	4401.00	14041.82	1150.31	870.16	4726.01	5369.07	1632.14	750.63
	2010	3839.6	13063.13	4581.92	14304.81	1251.6	905.27	4949.91	5667.84	1719.34	793.15
	2011	4195.03	13299.01	4572.66	14535.3	1342.96	941.79	5090.65	5897.61	1796.00	831.85
Simulating outcomes	2007	3128.55	13782.35	4235.96	15819.02	1136.21	1145.9	4971.62	5832.73	2030.99	719.11
	2008	3476.32	14033.49	4260.99	16031.54	1209.26	1188.26	5060.58	5928.71	2183.86	754.29
	2009	3838.23	14301.29	4281.13	16245.66	1286.92	1234.45	5154.92	6142.90	2350.95	788.71
	2010	4217.85	14572.06	4301.49	16465.88	1369.11	1282.07	5251.04	6364.03	2523.48	825.55
	2011	4618.25	14846.01	4322.48	16692.36	1456.22	1331.21	5349.38	6593.69	2702.68	864.96

Table 6.4 Calibration of current account balance percent of GDP (%)

		CHN	USA	JPN	EU	IND	RUS	ODC	HDC	MDC	LDC
IMF Data	2007	10.13	−5.06	4.82	−0.56	−0.70	5.93	3.15	4.03	0.74	2.25
	2008	9.12	−4.74	3.22	−1.02	−1.99	6.25	2.18	3.97	−1.01	0.14
	2009	5.23	−2.70	2.82	−0.09	−2.83	4.05	3.00	0.96	−0.32	−2.25
	2010	5.19	−3.24	3.59	−0.14	−2.62	4.81	3.04	1.67	−0.68	−0.44
	2011	5.16	−3.10	2.51	−0.18	−2.18	5.52	2.73	2.33	−1.00	1.10
Simulating outcomes	2007	9.90	−4.80	3.67	−0.71	−0.75	3.63	1.43	3.03	0.57	2.41
	2008	9.02	−4.75	3.78	−0.72	−1.10	3.54	1.59	3.12	0.01	1.80
	2009	8.17	−4.59	3.79	−0.70	−1.42	3.71	1.65	2.99	−0.52	1.47
	2010	7.38	−4.42	3.81	−0.67	−1.71	3.92	1.70	2.84	−0.99	1.19
	2011	6.66	−4.25	3.83	−0.64	−1.98	4.12	1.75	2.70	−1.43	0.93

0.9971; R^2 reaches 0.9941; P-value equals 0. In the Z-test, the Z value of these two sets of data is 0.47, lower than its one-tail threshold 1.64 and two-tail threshold 1.96. In ANOVA, the F value is 0.2248, lower than 1.6, the F-value upper bound in the case of the 50 × 50 sample size and 0.05 significance level. The calibration results demonstrate that the outcomes can reflect the economic growth trends exactly.

Table 6.4 presents the comparison of the Current Account (preset of GDP) of countries from 2007 to 2011 between the real data from IMF and the simulating results. The current account balance indicates the difference between a country's savings and its investment. If the current account balance is positive, it measures the portion of a country's saving invested abroad. The correlation coefficient between these two sets of data is 0.9273 in the regression analysis, and R^2 equals 0.86 and P-value equals 0. The Z value obtained in the Z-test is 0.0345, and the F-value from ANOVA is 0.0012. Although the correlation of the current account (present of GDP) is lower than that of GDP, it still higher than 0.9, and the results of the Z-test and ANOVA are both satisfactory. Thus, CIECIA is able to reflect the international capital flow well.

Table 6.5 presents the comparison of carbon emissions of countries from 2007 to 2011 between the real data from the EIA and the simulating results. The correlation coefficient between these two sets of data from the regression analysis is 0.9969, R^2 is 0.9939, and the P value is 0. Z from the Z-test is 0.0073 and F from ANOVA is 0. The Calibration results demonstrate that the carbon emissions of countries from the simulating outcomes agree with the real data well, depicting the trends of carbon emissions exactly under natural conditions.

6.4 Assessments of Global Cooperating Abatement Schemes

In the aspect of the assessment of the global abatement scheme, three principles, comprising effectiveness, feasibility and fairness should be followed for planning a reasonable global abatement scheme (Wang et al. 2014).

Table 6.5 Calibration of carbon emissions (MtC)

		CHN	USA	JPN	EU	IND	RUS	ODC	HDC	MDC	LDC
EIA Data	2007	1725.37	1643.53	342.12	1161.18	372.60	431.08	613.02	1150.05	527.10	143.01
	2008	1823.09	1593.98	331.71	1143.07	401.39	448.80	613.53	1197.25	552.39	146.42
	2009	2065.47	1482.35	301.34	1054.52	435.75	423.97	610.26	1171.02	560.04	141.44
	2010	2181.01	1537.29	321.76	1074.61	436.70	447.90	643.38	1230.35	574.05	144.49
	2011	2376.90	1497.44	321.99	1046.88	470.66	487.67	643.43	1307.98	585.00	147.10
Simulating outcomes	2007	1736.82	1847.02	1959.62	2069.73	2185.57	1736.82	1847.02	1959.62	2069.73	2185.57
	2008	1626.44	1612.77	1605.22	1586.51	1573.70	1626.44	1612.77	1605.22	1586.51	1573.70
	2009	339.22	331.64	325.76	318.44	310.34	339.22	331.64	325.76	318.44	310.34
	2010	1144.60	1129.95	1118.02	1115.05	1102.96	1144.60	1129.95	1118.02	1115.05	1102.96
	2011	375.82	394.29	412.54	432.09	453.15	375.82	394.29	412.54	432.09	453.15

The effectiveness of the abatement scheme refers to that the controlling target of the global mean surface temperature should be satisfied. The Copenhagen Consensus agrees to hold the increase in the global temperature below 2 °C before 2100. Meanwhile, according to Stern (2007), the atmospheric concentration equivalent of CO_2 should be controlled at the range of 450–500 ppmv. The feasibility refers to the possibility that the scheme could be accepted by the participating countries. From the perspective of climate ethics, it is believed that a feasible global abatement scheme needs to realize the Pareto improvement of the economic benefits of all of the participating countries, that is, the abatement scheme needs to guarantee that all of the participating countries benefit from carbon abatement activities, otherwise, global abatement cooperation will be impossible.

According to Santon et al. (2009), a global abatement scheme should consider not only its influences on developed countries but also the developments of developing countries, as well as the survivals of their people, i.e. the fairness of the scheme. It is difficult to determine the fairness of the scheme directly. To fully consider the fairness of the abatement schemes, this study divided the fairness into two types, the fairness of carbon emission permit allocation and the fairness of economic influences of the abatement scheme on countries.

In this section, the rising of the global mean surface temperature, and the economic developments and carbon emissions of countries in the non-abatement scenario (Scheme 0) are presented firstly. Then, six important global cooperating carbon abatement schemes are simulated, assessed and compared with each other based on Scheme 0. In this section, all of the abatement schemes are implemented from 2016, and the target years are 2050 and 2100.

6.4.1 The Non-Abatement Scheme

In Scheme 0, the global mean surface temperature increases by 3.16 °C in 2100 from the pre-industrial level, exceeding the 2 °C global warming target of the Copenhagen Consensus. In fact, the atmospheric concentration equivalent of CO_2 reaches 638 ppmv in 2100 under Scheme 0. Figure 6.1a, b present the accumulated utilities and the accumulated utilities per capital of countries respectively. It demonstrates that the accumulated utility of China increases fast, having exceeded the USA before 2075; the accumulated utilities from 2007 to 2100 of the USA, EU and HDC are similar to each other; Japan and Russia have lower accumulated utilities because of their smaller populations in future. In terms of per capita, all of the developed countries maintain high level of accumulated utilities per capital during the simulation; the accumulated utility per capital of China from 2007 to 2100 is only 251.23 thousand dollars (in constant 2005 prices), less than one half of the USA's accumulated utility per capital from 2007 to 2100; the accumulated utilities per capital of India and LDC are only 98.44 and 33.90 thousand dollars. Thus, in terms of accumulated utility per capita, the development of the world will still be imbalanced in the future.

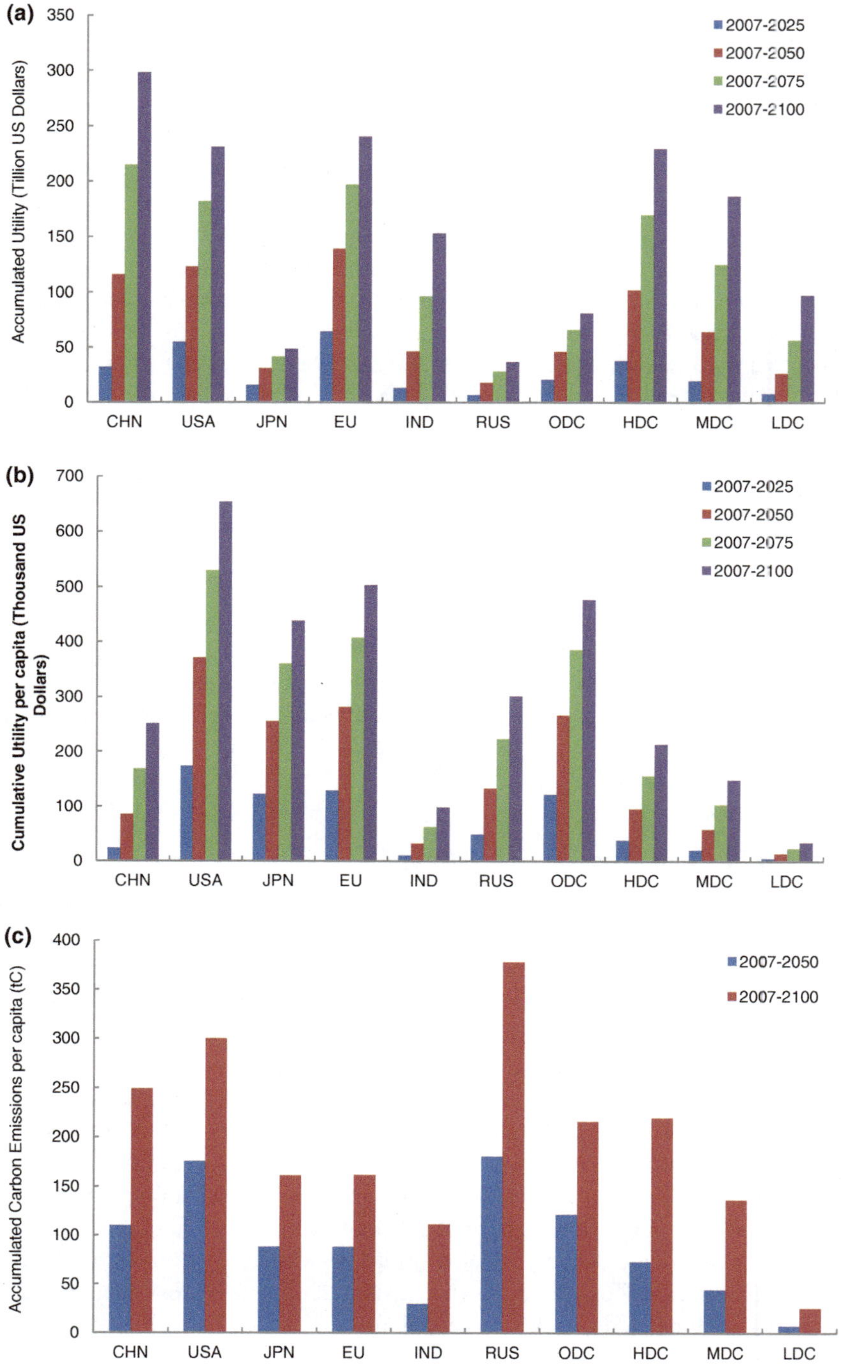
(a)
Accumulated Utility (Tillion US Dollars)
350
300
250
200
150
100
50
0
2007-2025
2007-2050
2007-2075
2007-2100
CHN
USA
JPN
EU
IND
RUS
ODC
HDC
MDC
LDC
(b)
Cumulative Utility per capita (Thousand US Dollars)
700
600
500
400
300
200
100
0
2007-2025
2007-2050
2007-2075
2007-2100
CHN
USA
JPN
EU
IND
RUS
ODC
HDC
MDC
LDC
(c)
Accumulated Carbon Emissions per capita (tC)
400
350
300
250
200
150
100
50
0
2007-2050
2007-2100
CHN
USA
JPN
EU
IND
RUS
ODC
HDC
MDC
LDC

◂**Fig. 6.1** **a** Accumulated utilities of countries in Scheme 0 (trillion US dollars in constant 2005 prices). **b** Accumulated utilities per capita of countries in Scheme 0 (thousand US dollars in constant 2005 prices). **c** Accumulated carbon emissions per capita of countries in Scheme 0 (tC)

Figure 6.1c depicts the accumulated carbon emissions per capita of countries from 2007 to 2050 and 2100 in Scheme 0. Among them, Russia has the largest accumulated carbon emissions per capita because of its relatively high output of the energy sector and low population numbers. The accumulated carbon emissions per capita of the USA from 2007 to 2100 are 300tC, lower than Russia's. The accumulated carbon emissions per capita of China by 2100 reach 250tC because of its relatively fast developing speed and high carbon intensity. The accumulated carbon emissions per capita of India and the LDC in Scheme 0 rank lowest because of their quite limited economic aggregates.

In the non-abatement scenario, the global carbon emissions increase continually during the simulation. In the year 2100, the global emissions of CO_2 will be 60.34GtCO_2. The changing trend of the global carbon emissions in Scheme 0 is located in the range of the baseline in IPCC-AR5 (IPCC 2014) under the assumption of the default growth, a bit lower than its average level.

6.4.2 Stern Scheme

Stern (2006) considered that the mitigation solution should be adopted immediately to reduce the emissions of carbon dioxide and other GHGs. The main target of Stern Scheme is 50% global carbon emission reductions by 2050, including 80% reductions by the developed countries by 2050. In this section, the Stern Scheme is completed and simulated based on CIECIA. The abatement targets of countries in different periods are listed in Table 6.6. For convenience, the Stern Scheme is referred to as Scheme 1.

In Scheme 1, the global carbon emissions in 2100 increase to 2.90GtC, meeting the target of 50% reduction. The global mean surface temperature increases by 1.82 °C from the pre-industrial level and the atmospheric concentration equivalent of CO_2 reaches 444.51 ppmv in 2100, meeting both the global warming target and the GHG concentration target. Thus, Scheme 1 is an effective global cooperating abatement scheme.

Figure 6.2a depicts the accumulated carbon emissions per capital of countries in Scheme 1. It can be seen that although the reduction targets of the developed countries are higher than those of the developing countries, the reduction rate of accumulated carbon emissions per capital by 2100 of developing countries e.g. China and India are much higher than those of the developed countries. The accumulated carbon emissions per capital of India decrease from 111.22tC in Scheme 0 to 82.17tC in Scheme 1, whereas the reduction rates of the USA and Japan are lower than 50%. That is mainly because that the emissions of developed

Table 6.6 Reduction targets of countries in Scheme 1

	By 2050	By 2100	Carbon permits 2007–2050 (GtC)	Carbon permits 2007–2100 (GtC)
CHN	50% reduction based on 2005 level	50% reduction based on 2050 level	77.75	105.50
USA	80% reduction based on 1990 level	50% reduction based on 2050 level	44.68	54.92
JPN	80% reduction based on 1990 level	50% reduction based on 2050 level	8.50	10.36
EU	80% reduction based on 1990 level	50% reduction based on 2050 level	31.87	39.92
IND	30% reduction based on 2005 level	50% reduction based on 2050 level	17.41	25.81
RUS	70% reduction based on 1990 level	50% reduction based on 2050 level	15.73	21.88
ODC	80% reduction based on 1990 level	50% reduction based on 2050 level	17.08	19.81
HDC	60% reduction based on 1990 level	50% reduction based on 2050 level	39.89	52.93
MDC	30% reduction based on 2005 level	50% reduction based on 2050 level	24.22	37.17
LDC	Start reduction from 2040; maintain 2040's emission level between 2041 and 2050	Maintain 2050's reduction rate	13.66	60.26

countries present a downward trend from approximately 2010, whereas the energy demands of developing countries such as China will still increase after 2010 because of their industrializations. This demonstrates that Scheme 1 is unfair in the assignment of reduction tasks of countries. Considering that China, India and Russia are the main manufacturing forces in the world, this economic shock may have negative effect on the global economy, because it is hard to image a world lacking in manufacturing.

Figure 6.2b depicts the changing rates of the accumulated utility of the world and the countries from Scheme 0 to Scheme 1. Influenced by the implementations of this abatement scheme, the accumulated utilities of countries suffer before 2050 except the LDC having no abatement task before 2040. However, along with the

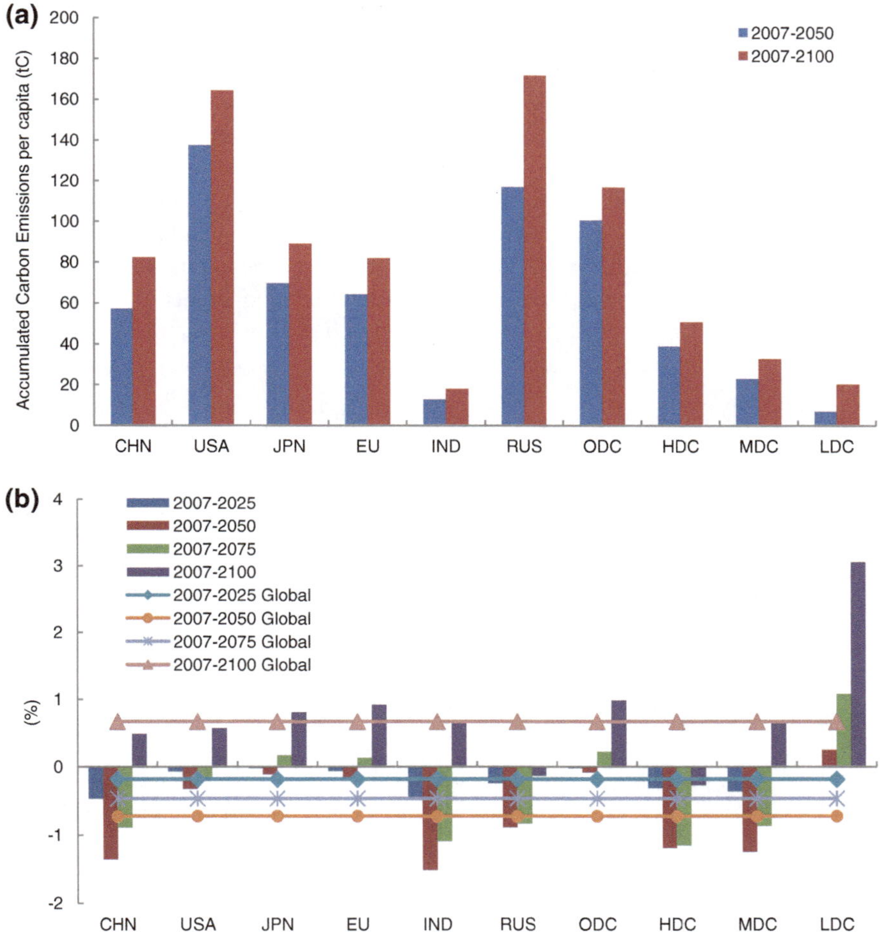

Fig. 6.2 **a** Accumulated carbon emissions per capita of countries in Scheme 1 (tC). **b** Changing rates of accumulated utility in Scheme 1 (%)

control of global warming after 2050, the accumulated utilities of countries improve from the same steps in Scheme 0 by the climate welfare. In Scheme 1, the improve rates of developed countries are higher than those of developing countries obviously, which means that Scheme 1 is unfair to the developing countries and the developed countries benefit more from Scheme 1.

6.4.3 *Norhaus Scheme*

Because of the unfairness and unreasonableness of the Stern Scheme, Nordhaus (2008) proposed that global carbon emissions should be reduced step by step, instead of the radical abatement schemes. According to the assumption of Nordhaus, the world has a target of 25% global carbon emission reductions by 2050. The developed countries have a target of 80% reductions by 2050, the same as Scheme 1, and the developing countries cut their emissions appropriately. The reduction targets of countries in Nordhaus Scheme (Scheme 2 for short) are listed in Table 6.7.

In Scheme 2, the global carbon emissions increase to 4.55GtC in 2050, meeting the abatement target of 25% global reduction by 2050. In 2100, global mean surface temperature increases by 1.98 °C, and the atmospheric concentration equivalent of CO_2 reaches to 462.51 ppmv, meeting the climate mitigation targets by 2100. Thus, Scheme 2 is also an effective global cooperating abatement scheme.

Table 6.7 Reduction targets of countries in Scheme 2

	By 2050	By 2100	Carbon permits 2007–2050 (GtC)	Carbon permits 2007–2100 (GtC)
CHN	Reduce to 2006 level by 2030; Maintain 2006's emission level from 2031o 2050	50% reduction based on 2050 level	83.36	143.63
USA	80% reduction based on 1990 level	50% reduction based on 2050 level	44.68	54.92
JPN	80% reduction based on 1990 level	50% reduction based on 2050 level	8.50	10.36
EU	80% reduction based on 1990 level	50% reduction based on 2050 level	31.87	39.92
IND	Maintain 2015's emission level	50% reduction based on 2050 level	23.15	43.42
RUS	70% reduction based on 1990 level	50% reduction based on 2050 level	15.73	21.88
ODC	80% reduction based on 1990 level	50% reduction based on 2050 level	17.08	19.81
HDC	60% reduction based on 1990 level	50% reduction based on 2050 level	41.47	57.76
MDC	Maintain 2015's emission level	50% reduction based on 2050 level	30.99	57.98
LDC	Start reduction from 2040; maintain 2040's emission level between 2041 and 2050	Maintain 2050's reduction rate	13.67	60.08

Figure 6.3a presents the accumulated carbon emissions per capita of countries in Scheme 2. Under this scheme, the accumulated carbon emissions per capita of developing countries increase obviously by 2100 compared with those in Scheme 1. Among them, the 2007–2100 accumulated carbon emissions per capita of China reach 115.76tC, having a one-half reduction compared with that of Scheme 0. The accumulated carbon emissions per capita of India and the MDC from 2007 to 2100 increase to 29.02 and 48.96tC respectively, both higher than those of the Stern Scheme. That means the developing countries obtain more carbon permits in Scheme 2; thus, Scheme 2 benefits the developing countries more.

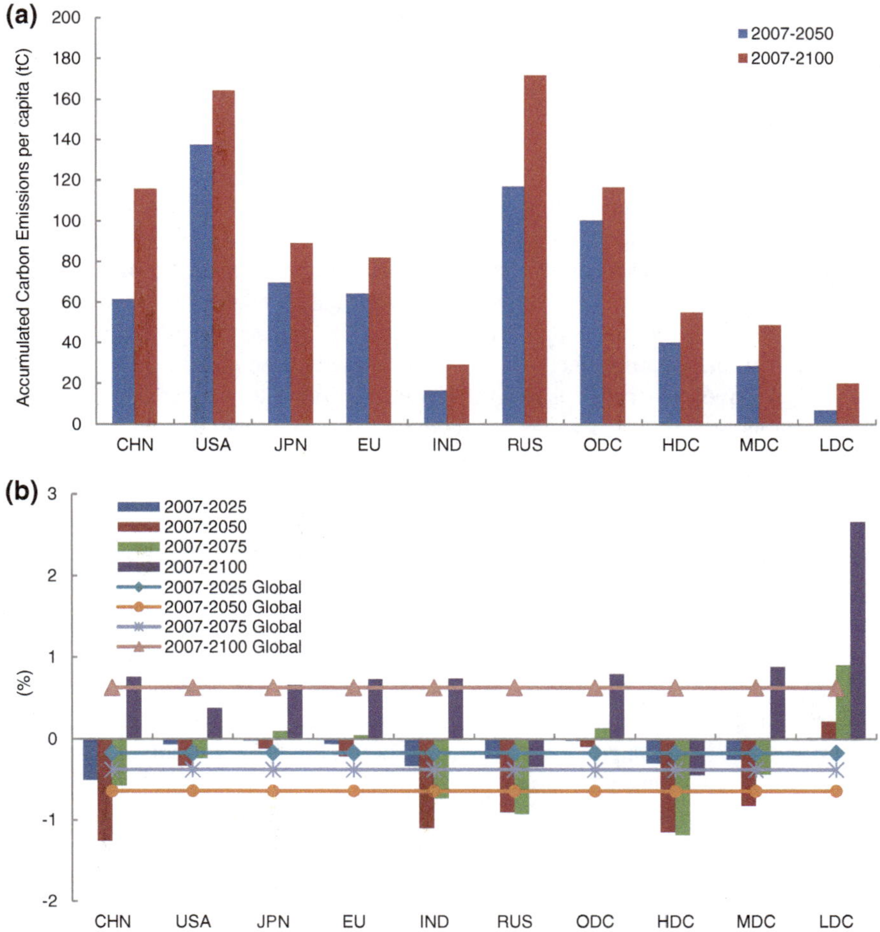

Fig. 6.3 **a** Accumulated carbon emissions per capita of countries in Scheme 2 (tC). **b** Changing rates of accumulated utility in Scheme 2 (%)

Figure 6.3b presents the rates of change in the accumulated utility of the world and countries from Scheme 0 to Scheme 2. The increasing rates of 2007–2100 accumulated utility of developing countries are higher than developed countries, which further indicates that Scheme 2 benefits the economies of the developing countries more. The increasing rate of accumulated utility of the MDC from 2007 to 2100 is 0.88%, in the second place in the ranking; China has the third highest increasing rate of accumulated utility by 2100, after LDC and MDC.

However, the changing rates of the accumulated utility of Russia by 2100 are both negative in Schemes 1 and 2, which means that these two abatement schemes will cause losses for the economy of Russia which cannot be offset only by the climate welfare. Thus, Schemes 1 and 2 are not feasible and cannot be accepted by all of the participating countries.

6.4.4 Principle of Convergence on Carbon Emissions Per Capita

Many scholars have proposed that fairness for emission reduction is to converge the carbon emissions per capita. He (2004) considered that equality of global per capita emissions should be taken as the standard for the fairness principle of the carbon permit allocation of countries. The principle of per capita reflects the equal rights of human survival, development and usage of natural resources. Based on the principle of abatement task sharing, the basic idea of the carbon emissions per capita convergence scheme (Scheme 3) is that the developed countries cut their emissions per capita gradually from 2016 while the developed countries increase their emissions per capita, and the carbon emissions per capita of all of the countries converges to a unique value by the target year. In consideration of the abatement restrict in 2050, the detailed global abatement scheme is listed in Table 6.8.

The global carbon emissions in 2050 increase to 3.63GtC in Scheme 3, between Schemes 1 and 2. In 2100, the global mean surface temperature increases by 1.98 °C, and the atmospheric concentration equivalent of CO_2 is 469.04 ppmv, meeting the climate mitigation targets by 2100.

Figure 6.4a demonstrates that the carbon emissions per capita of all of the countries, including LDC, converge to 0.5tC in 2100, realizing the convergence of carbon emissions per capita by the target year. Figure 6.4b presents the accumulated carbon emissions per capita of countries. In Scheme 3, the accumulated carbon emissions per capita of India, HDC, MDC and LDC are all higher than those in Schemes 1 and 2, whereas the accumulated carbon emissions per capita of China and Russia are lower than those in Scheme 2. That means China benefits less from Scheme 3 in the actual conditions and the developing trends.

Figure 6.4c depicts the changing rates of accumulated utility of the world and the countries in Scheme 3 relative to Scheme 0. The accumulated utilities of China and Russia decline obviously compared with those in Schemes 1 and 2. The

Table 6.8 Reduction targets of countries in Scheme 3

	By 2050	By 2100	Carbon permits 2007–2050 (GtC)	Carbon permits 2007–2100 (GtC)
CHN	68% reduction based on 2005 level	Maintain 2050's emission level	72.89	96.62
USA	85% reduction based on 1990 level	Maintain 2050's emission level	43.39	53.56
JPN	82% reduction based on 1990 level	Maintain 2050's emission level	8.40	10.62
EU	79% reduction based on 1990 level	Maintain 2050's emission level	32.03	43.27
IND	Start reduction from 2023; maintain 2023's emission level between 2024 and 2050	Maintain 2050's emission level	29.55	66.88
RUS	94% reduction based on 1990 level	Maintain 2050's emission level	13.72	16.38
ODC	80% reduction based on 1990 level	Maintain 2050's emission level	17.24	21.33
HDC	43% reduction based on 1990 level	Maintain 2050's emission level	42.62	67.69
MDC	Reduce to 2012's emission level	Maintain 2050's emission	29.84	62.87
LDC	No abatement task	Start reduction from 2096; maintain 2096's emission level between 2097 and 2100	14.67	81.45

accumulated utility changing rate of China is estimated to be only 0.02% by 2100, and the changing rate of Russia decrease to −0.86%. India can benefit most in Scheme 3. The changing rate of India's accumulated utility reaches to 1.44%, much higher than those of most of the other countries.

Those countries that have large population bases and relatively low capital emission demands per capita will benefit from Scheme 3. In addition, developed countries with higher decrease rates of carbon intensity e.g. the EU and Japan will not suffer great loss in this scheme. However, countries that have higher initial capital emissions per capita and faster decline rates of population (according to UNDESA, China will meet its population peak approximately 2030) will lose their development chances. Scheme 3 avoids the historic responsibilities of the developed countries in the carbon emission problem, unable to control the carbon emissions of the developed countries well. Instead, it brings harm to important economies such as China and Russia. Thus, Scheme 3 is not feasible.

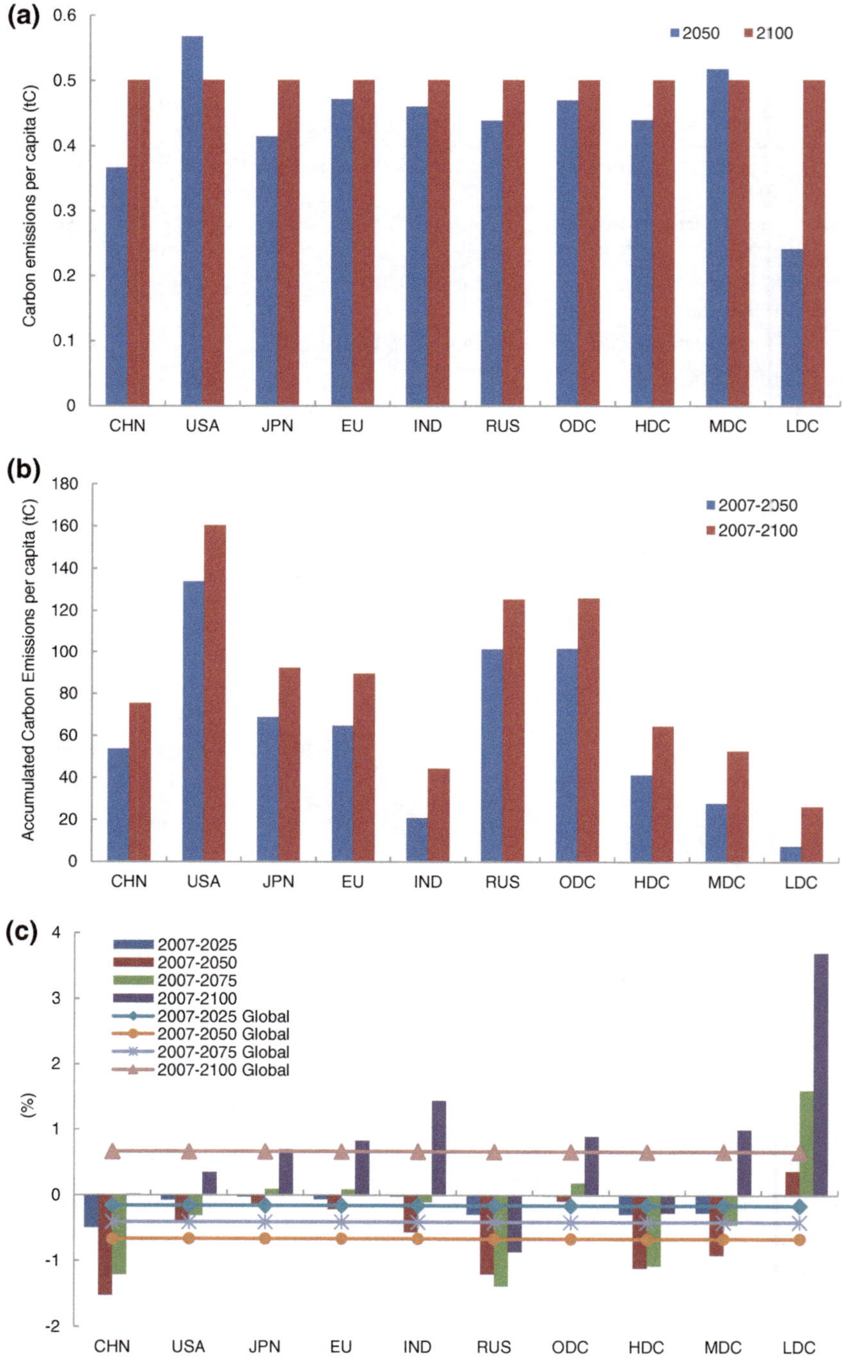

Fig. 6.4 **a** Carbon emissions per capita of countries in Scheme 3 (tC). **b** Accumulated carbon emissions per capita of countries in Scheme 3 (tC). **c** Changing rates of accumulated utility in Scheme 3 (%)

6.4.5 *Principle of Convergence on Accumulated Carbon Emissions Per Capita*

The principle of convergence on accumulated carbon emissions per capita, in which the accumulated carbon emissions per capita of countries are converged by the target year 2100, was primarily proposed by scholars from developing countries including Ding et al. (2009). It emphasizes the fairness of the carbon abatement from the perspective of historic carbon emissions. Under the principle of convergence on accumulated carbon emissions per capita, the developed countries will have heavy reduction tasks in the short or medium term, for realizing the emission routs under the global long-term reduction target and setting apart necessary development spaces for the developing countries. In this study, three accumulated carbon emissions per capita convergence schemes with different start years of emission accumulation (2007, 2000 and 1990) are assessed: Schemes 4a, 4b and 4c. The target year of these three schemes are all 2100. The detailed reduction targets of countries are listed in Table 6.9.

In these three schemes, the global mean surface temperature in 2100 increases by 1.98 °C in Scheme 4a, 1.99 °C in Scheme 4b and 1.99 °C in Scheme 4c respectively. The atmospheric concentration equivalent of CO_2 in 2100 is 465.88 ppmv in Scheme 4a, 472.01 ppmv in Scheme 4b and 474.50 ppmv in Scheme 4c. All of these three schemes are effective for meeting the climate mitigation targets by 2100.

Figure 6.5a–c present the accumulated carbon emissions per capital of countries in these three schemes respectively. In Scheme 4a, the accumulated carbon emissions per capital of countries from 2007 to 2100 all reach 67.5tC except LDC. In Scheme 4b, the accumulated carbon emissions per capital of countries from 2000 to 2100 converge to 78tC except for the USA and LDC. Although the USA realizes 100% reductions from 2016 in Scheme 4b, its accumulated emissions per capital of countries from 2000 to 2100 are still higher than those of the other countries. In Scheme 4c, the USA and ODC achieve 100% reductions from 2016; Japan and the EU should reach 100% reductions before 2050; the accumulated carbon emissions per capita of the other countries except LDC converge to 87tC. For the developed countries, Schemes 4b and 4c require 100% reductions in a short time, which cause difficulties for the implementation of the schemes.

Figure 6.6a–c depict the changing rates of the accumulated utility of the world and countries in the three schemes, respectively, relative to Scheme 0. In Scheme 4a, India, HDC and MDC have the highest changing rates compared with those in Schemes 1, 2 and 3. The changing rates of the accumulated utility of India is 2.71% by 2100. Although the changing rate of HDC by 2100 is negative, it is still much higher than those in the schemes assessed previously. The changing rates of the accumulated utility of China, Russia and the developed countries in Scheme 4a decrease obviously. The changing rate of the accumulated utility of China is less than 0.01% by 2100, and the USA's even turns to the negative.

Along with the moving of the start year of accumulation to 2000, the accumulated utilities of India and MDC increase. In Scheme 4c, the changing rate of the

Table 6.9 Reduction targets of countries in Scheme 4

Scheme	Country/region	By 2050	By 2100	Carbon permits 2007–2050 (GtC)	Carbon permits 2007–2100 (GtC)
Scheme 4a	CHN	71% reduction based on 2005 level	50% reduction based on 2050 level	72.08	88.10
	USA	95% reduction based on 1990 level by 2020; maintain 2020's emission level between 2021 and 2050	50% reduction based on 2050 level	19.19	21.50
	JPN	96% reduction based on 1990 level	50% reduction based on 2050 level	7.79	8.18
	EU	91% reduction based on 1990 level	50% reduction based on 2050 level	29.69	33.23
	IND	Start reduction from 2048; maintain 2048's emission level between 2049 and 2050	50% reduction based on 2050 level	43.37	104.95
	RUS	97% reduction based on 1990 level by 2030; maintain 2030's emission level between 2031 and 2050	50% reduction based on 2050 level	8.36	9.18
	ODC	94% reduction based on 1990 level by 2030; maintain 2030's emission level between 2031 and 2050	50% reduction based on 2050 level	10.41	11.27
	HDC	37% reduction based on 2005 level	50% reduction based on 2050 level	45.83	71.16
	MDC	Start reduction from 2028; maintain 2028's emission level between 2029 and 2050	50% reduction based on 2050 level	41.00	81.10
	LDC	No abatement task	No abatement task	/	/

(continued)

Table 6.9 (continued)

Scheme	Country/region	By 2050	By 2100	Carbon permits 2007–2050 (GtC)	Carbon permits 2007–2100 (GtC)
Scheme 4b	CHN	65% reduction based on 2005 level	50% reduction based on 2050 level	73.79	93.35
	USA	100% reduction from 2016	50% reduction based on 2050 level	14.13	14.13
	JPN	90% reduction based on 1990 level by 2035; maintain 2035's emission level between 2036 and 2050	50% reduction based on 2050 level	6.16	7.17
	EU	96% reduction based on 1990 level	50% reduction based on 2050 level	28.87	30.73
	IND	Start reduction from 2043; maintain 2043's emission level between 2044 and 2050	Maintain 2050's emission level	42.42	118.92
	RUS	99% reduction based on 1990 level by 2030; maintain 2030's emission level between 2031 and 2050	50% reduction based on 2050 level	7.90	8.10
	ODC	93% reduction based on 1990 level by 2020; maintain 2020's emission level between 2021 and 2050	50% reduction based on 2050 level	7.66	8.67
	HDC	30% reduction based on 2005 level	50% reduction based on 2050 level	47.09	75.03
	MDC	Start reduction from 2033; maintain 2033's emission level between 2034 and 2050	50% reduction based on 2050 level	44.09	90.15
	LDC	No abatement task	No abatement task	/	/

(continued)

Table 6.9 (continued)

Scheme	Country/region	By 2050	By 2100	Carbon permits 2007–2050 (GtC)	Carbon permits 2007–2100 (GtC)
Scheme 4c	CHN	61% reduction based on 2005 level	50% reduction based on 2050 level	74.78	96.39
	USA	100% reduction from 2016	50% reduction based on 2050 level	14.13	14.13
	JPN	100% reduction from 2036	50% reduction based on 2050 level	5.63	5.63
	EU	100% reduction from 2042	50% reduction based on 2050 level	23.74	23.74
	IND	Start reduction from 2049; maintain 2049's emission level in 2050	maintain 2050's emission level	43.05	129.56
	RUS	100% reduction from 2018	50% reduction based on 2050 level	4.73	4.73
	ODC	100% reduction from 2016	50% reduction based on 2050 level	5.57	5.57
	HDC	35% reduction based on 2005 level	50% reduction based on 2050 level	46.19	72.27
	MDC	Start reduction from 2028; maintain 2028's emission level between 2029 and 2050	Maintain 2050's emission level	40.96	94.79
	LDC	No abatement task	No abatement task	/	/

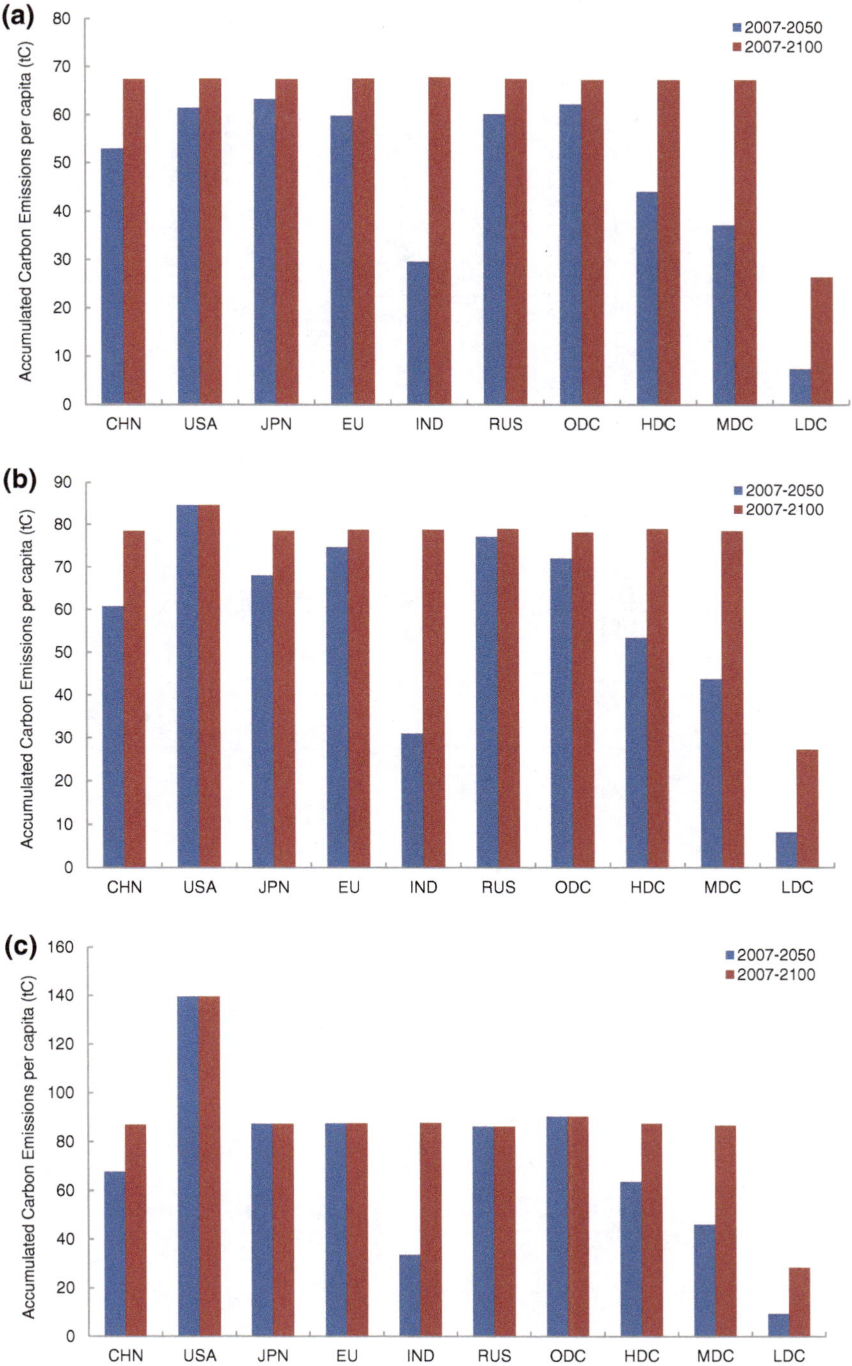

Fig. 6.5 **a** Accumulated carbon emissions per capita of countries in Scheme 4a (tC). **b** Accumulated carbon emissions per capita of countries in Scheme 3 (tC). **c** Accumulated carbon emissions per capita of countries in Scheme 4c (tC)

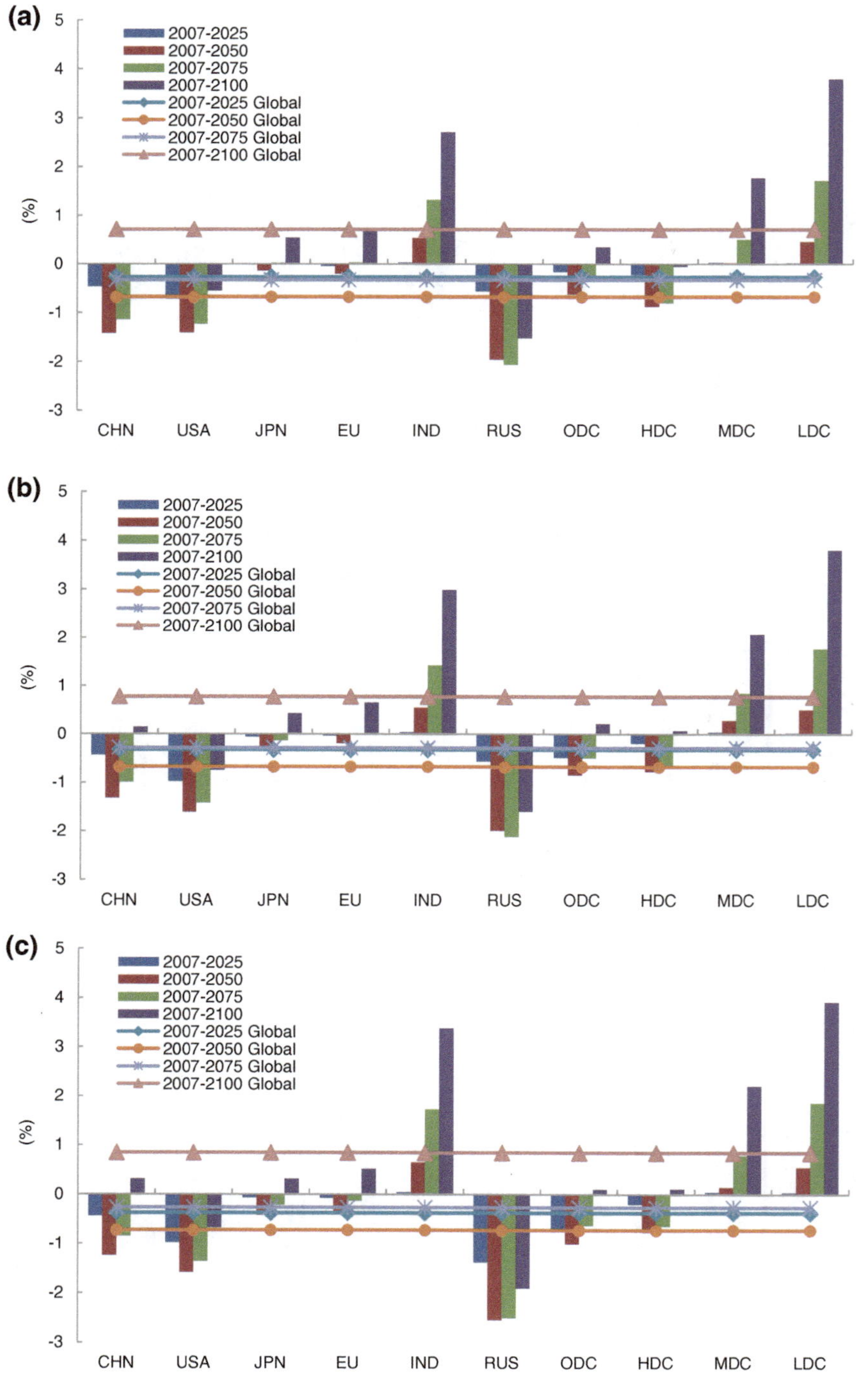

Fig. 6.6 **a** Changing rates of accumulated utility in Scheme 4a (%). **b** Changing rates of accumulated utility in Scheme 4b (%). **c** Changing rates of accumulated utility in Scheme 4c (%)

accumulated utility of India by 2100 reaches to 3.38%, close to LDC which has no abatement task, and the changing rate of MDC by 2100 increases to 2.20%. Meanwhile, the changing rates of the accumulated utility of the developed countries decrease generally. The changing rate of the EU by 2100 is only 0.51%, below the global average level of 0.87% in Scheme 4c. The changing rates of the accumulated utility of China and HDC increase slightly compared with Scheme 4a, along with the moving of the start year.

From the perspective of global economic growth, the changing rates of the global accumulated utility by 2025 are −0.27%, −0.34% and −0.38%, respectively, in Schemes 4a, 4b and 4c, much lower than the other schemes. That indicates that the carbon emission controls of the main economies in the early steps of Scheme 4 are overly strict so that the global economy suffers large losses and declines significantly.

Scheme 4 benefits those countries with huge population bases and increasing population trends, whereas the emissions of the developed countries are controlled severely and their economic developments are affected seriously. Scheme 4 may hurt the benefit of the global economy, resulting in economic setbacks for the world. Thus, the feasibility of Scheme 4 is still questionable. In addition, although the scholars of China support the principle of convergence on accumulated carbon emissions per capita in the international climate conferences, China will not achieve the development chances under this principle in the long run, even the star year of accumulation is 1990, primarily because the population of China will decrease from approximately 2030.

6.4.6 Global Economic Growth Scheme

To avoid the serious damages to the main economies of the world, Wang et al. (2012) proposed that the carbon abatements should consider the steady growth of the global economy. Both China and the USA are the main manufacturing countries and the economic engines of the world. The overly stringent reduction tasks for China and the USA may lead to global economic disasters. Thus, the carbon emissions per capita of China and the USA should be allowed to be slightly higher than those in other schemes. Meanwhile, the abatement tasks of the middle and low developing countries should be reduced as little as possible, giving them adequate development opportunities. Table 6.10 lists the abatement targets of countries in the economic growth scheme (Scheme 5). The start year of abatement of LDC is postponed until 2045 in Scheme 5.

In Scheme 5, the global carbon emission reaches to 3.88GtC in 2050, close to that of Scheme 3. By 2100, the global mean surface temperature increases by 1.99 °C, and the atmospheric concentration equivalent of CO_2 is 467.26 ppmv, meeting the climate mitigation targets by 2100.

Figure 6.7a depicts the accumulated carbon emissions per capita of countries in Scheme 5. The accumulated carbon emissions per capita of China is 111.88tC by

Table 6.10 Reduction targets of countries in Scheme 5

	By 2050	By 2100	Carbon permits 2007–2050 (GtC)	Carbon permits 2007–2100 (GtC)
CHN	30% reduction based on 2005 level	Maintain 2050's emission level	87.08	139.23
USA	80% reduction based on 1990 level	20% reduction based on 2050 level	44.68	57.03
JPN	80% reduction based on 1990 level	20% reduction based on 2050 level	8.50	10.74
EU	80% reduction based on 1990 level	20% reduction based on 2050 level	31.87	41.57
IND	Reduce to 2010 level	Maintain 2050's emission level	21.13	42.74
RUS	60% reduction based on 1990 level	40% reduction based on 2050 level	16.72	25.48
ODC	80% reduction based on 1990 level	20% reduction based on 2050 level	17.08	20.37
HDC	50% reduction based on 1990 level	30% reduction based on 2050 level	41.47	59.99
MDC	Reduce to 2010 level	30% reduction based on 2050 level	28.79	54.28
LDC	Start reduction from 2045; maintain 2045's emission level between 2046 and 2050	Maintain 2050's emission level	14.31	69.25

2100, close to those in Scheme 2. The accumulated carbon emissions per capita of the developed countries, e.g., the USA and EU, increase obviously compared with those in Scheme 2. In addition, the accumulated carbon emissions per capita of Russia by 2100 increase to 203.12tC.

Figure 6.7b depicts the changing rates of accumulated utility of the world and countries in Scheme 5 relative to Scheme 0. In Scheme 5, the changing rates of accumulated utility are relatively equal among countries. China, India, MDC, the EU, Japan and the ODC have almost the same changing rate by 2100. The changing rate of Russia by 2100 is close to 0, higher than those changing rates in the other schemes. However, the changing rate of accumulated utility of HDC by 2100 is still negative in Scheme 5; thus, Scheme 5 still may not be accepted by all countries.

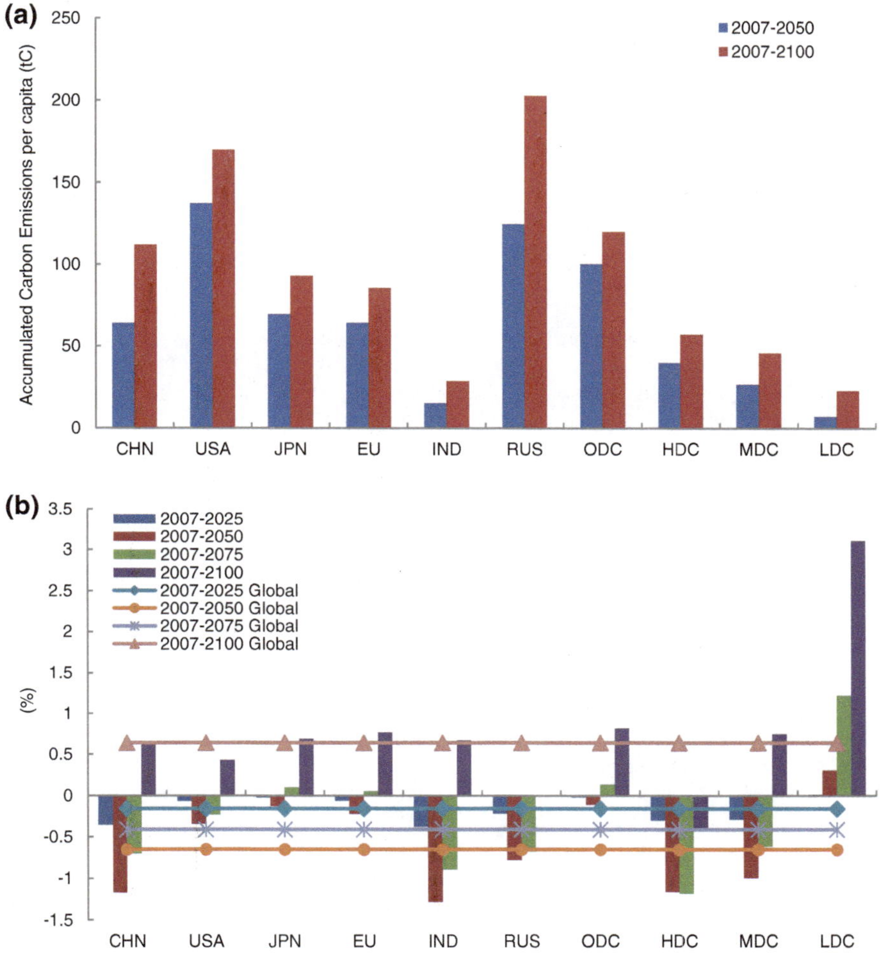

Fig. 6.7 **a** Accumulated carbon emissions per capita of countries in Scheme 5 (tC). **b** Changing rates of accumulated utility in Scheme 5 (%)

6.4.7 Pareto Improvement Scheme

From the perspective of welfare economics, Wang et al. (2014) noted that a global abatement scheme will not be accepted by all of the cooperating countries unless it can benefit all of the abatement participators. Such a type of scheme that can improve the welfares of the participators is named the Pareto Improvement Scheme. Pareto improvement refers to a type of social change to improve the social welfare of some without harming the welfares of the others. The target of the Pareto Improvement Scheme is to guarantee that the changing rates of the utility of all of the countries are positive by 2100 on the premise that the climate mitigation targets

Table 6.11 Reduction targets of countries in Scheme 6

	By 2050	By 2100	Carbon permits 2007–2050 (GtC)	Carbon permits 2007–2100 (GtC)
CHN	25% reduction based on 2005 level	40% reduction based on 2050 level	84.46	128.93
USA	75% reduction based on 1990 level	40% reduction based on 2050 level	45.92	59.60
JPN	90% reduction based on 1990 level	40% reduction based on 2050 level	8.05	9.04
EU	95% reduction based on 1990 level	40% reduction based on 2050 level	28.96	31.11
IND	Reduce to 2005 level	20% reduction based on 2050 level	19.15	33.61
RUS	50% reduction based on 1990 level	40% reduction based on 2050 level	17.71	28.65
ODC	98% reduction based on 1990 level	70% reduction based on 2050 level	15.90	16.13
HDC	20% reduction based on 2005 level	20% reduction based on 2050 level	49.08	87.72
MDC	10% reduction based on 2005 level	20% reduction based on 2050 level	26.00	46.09
LDC	Start reduction from 2045; maintain 2045's emission level between 2046 and 2050	Maintain 2050's emission level	14.30	69.23

by 2100 would be achieved. The detailed reduction targets of the Pareto Improvement Scheme (Scheme 6) are listed in Table 6.11.

In Scheme 6, the global mean surface temperature increases by 1.97 °C in 2100 compared with the pre-industrial level, and the atmospheric concentration equivalent of CO_2 is 464.17 ppmv, meeting the climate mitigation targets by 2100. Thus, Scheme 6 is effective. Figure 6.8a presents the accumulated carbon emissions per capital of countries. The accumulated carbon emissions per capital from 2007 to 2100 of Russia and HDC increase to 230.49tC and 82.69tC respectively, and ODC with higher changing rates of accumulated utility in Scheme 5 has an even more stringent reduction task after 2050 in Scheme 6. That is the sacrificed part of the fairness for Pareto improvement.

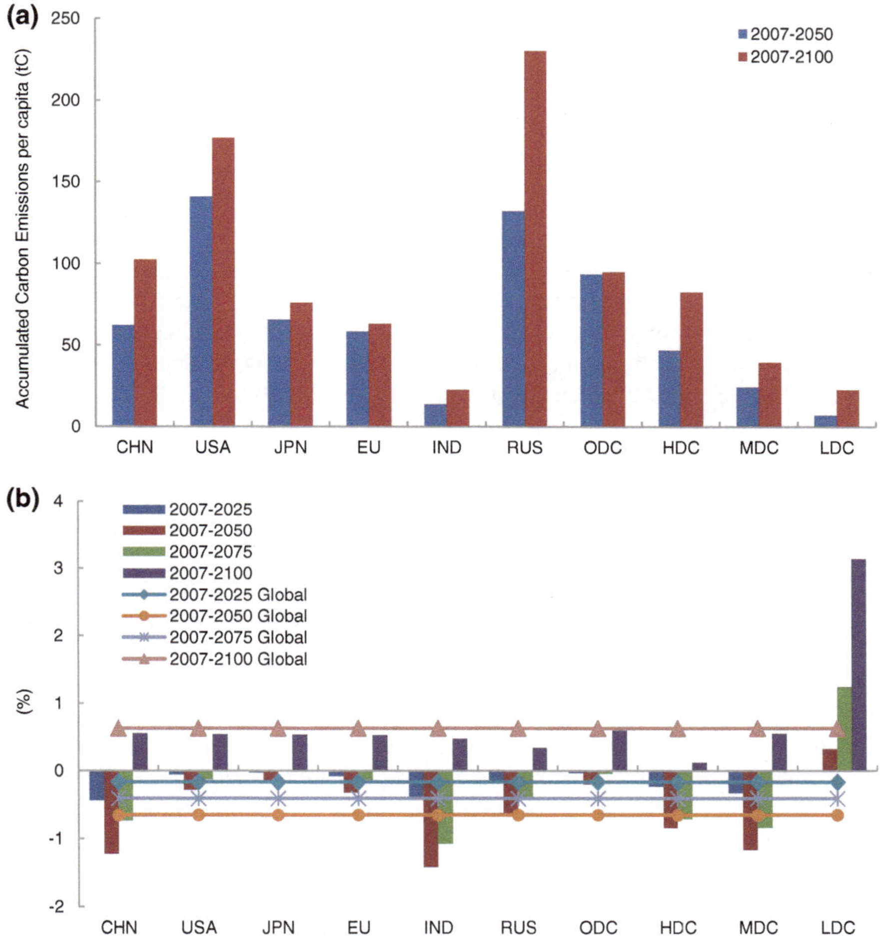

Fig. 6.8 **a** Accumulated carbon emissions per capita of countries in Scheme 6 (tC). **b** Changing rates of accumulated utility in Scheme 6 (%)

Figure 6.8b depicts the changing rates of accumulated utility of the world and countries in Scheme 6 relative to Scheme 0. It demonstrates that the accumulated utility of all of the countries by 2100 are higher than those in Scheme 0, which means the Pareto improvement of carbon abatement is realized in Scheme 6. Meanwhile, except HDC and LDC, the changing rates of countries by 2100 are almost equal. Thus, Scheme 6 can provide good fairness among countries in the aspect of the economic influence of carbon abatement.

Schemes 5 and 6 are designed and improved based on Schemes 1 and 2. Compared with Schemes 1 and 2, Schemes 5 and 6 consider the steady economic growth of the world and development demands of the developing countries comprehensively and is more reasonable. Scheme 6 realizes the Pareto improvement of accumulated utilities of all of the abatement participators, making it acceptable to all countries. Thus, Scheme 6 is the most feasible global cooperating abatement scheme at present.

Table 6.12 presents the effectiveness, fairness and feasibilities of schemes. In the aspect of effectiveness, global warming by 2100 in the schemes are all below 2 °C, and the atmospheric concentration equivalent of CO_2 by 2100 are all controlled below 500 ppmv. In the aspect of fairness, Schemes 3 and 4 with the principles of convergence on carbon emissions per capita and convergence on accumulated carbon emissions per capita respectively have the lowest Gini coefficients. Schemes 5 and 6 have the third and fourth lowest Gini coefficients after Scheme 3 and 4. In the aspect of feasibility, the main economies, e.g., China, the USA and Russia, suffer seriously under Schemes 3 and 4, which makes neither Scheme 3 nor Scheme 4 feasible. The accumulated utility of HDC by 2100 in Scheme 5 is lower than that in Scheme 0, which greatly reduces the feasibility of Scheme 5. The Pareto improvement scheme is the only one that can improve the benefits of all of the abatement participating countries and is thus most feasible. In summary, although sacrificing some fairness, the Pareto Improvement Scheme, which was developed from Scheme 5, is the most feasible scheme and is thus the most reasonable global cooperating climate mitigation scheme.

Comparing the abatement schemes with the emission scenarios in IPCC-AR5, the global carbon emissions of Schemes 1, 2, 3, 5 and 6 by 2020 are all located around the lower bound of the range for the Cancun Agreement (30–33 $GtCO_2$). In Scheme 4, because of huge emission cuts of the developed countries in the initial steps, the global carbon emissions in 2020 are relatively small, generally between 30 and 33$GtCO_2$. In general, the global carbon emissions from 2010 to 2030 of schemes in this study are close to the outcomes of OS/No Negative/Full scenario categories (Categories 0–1) for meeting a goal of 450 ppmv by the end of the century. From a long-term perspective, the global mean surface temperature in the schemes increases between 1.8 and 2 °C in 2100, and the atmospheric concentration equivalent of CO_2 by 2100 is located between 400 ppm and 435 ppm; the global accumulated carbon emissions from 2010 to 2100 are between 1900 and 2100GtC. These outcomes are quite close to the Scenario Category 2 of IPCC-AR5. Among them, the Stern Scheme is much closer to Category 1. In addition, the abatement routs of the schemes in this study are quite close to those in the No Negative scenarios of Category 1, i.e., huge emission cuts in the early steps and gradually bringing down the mitigation rates in the later stages.

The changing rates of the global utility of schemes relative to Scheme 0 are summarized in Fig. 6.9. It is easy to find that the scheme under the principle of convergence on accumulated carbon emissions per capita (Scheme 4) obtains the highest global accumulated utility from 2007 to 2100. However, this scheme hurts the interests of the main economies in the world and has negative impacts on global economic development in the short term. Thus, in the currently depressed economic conditions, Scheme 4 is deprecated; although from a long-term perspective, this scheme would benefit middle and low developing countries more and promote the global economic development in the later period. This problem is worthy of further research.

Table 6.12 Comparison of effectiveness, fairness and feasibility of schemes

	Effectiveness		Fairness		Feasibility
	Global warming by 2100 (°C)	Atmospheric concentration equivalent of CO_2 by 2100	Gini coefficient of carbon emissions in 2100	Gini coefficient of accumulated carbon emissions between 1990 and 2100	
Scheme 1	1.8213	444.51	0.2558	0.4078	Siding with the developed countries, infeasible
Scheme 2	1.9842	462.69	0.2383	0.3857	The changing rates of accumulated utility of Russia and HDC by 2100 are negative, low feasibility
Scheme 3	1.9789	469.04	0.0000	0.3005	The interests of main economies, e.g., China and Russia, are damaged, infeasible
Scheme 4a	1.9760	465.88	0.198	0.1974	The interests of main economies, e.g., China, the USA and Russia, are damaged, infeasible
Scheme 4b	1.9998	472.01	0.2809	0.1813	The interests of main economies, e.g., China, the USA and Russia, are damaged, infeasible
Scheme 4c	1.9903	474.50	0.3207	0.1934	The interests of main economies, e.g., China, the USA and Russia, are damaged, infeasible
Scheme 5	1.9884	467.26	0.2180	0.3760	The changing rates of accumulated utility of HDC by 2100 are negative, low feasibility
Scheme 6	1.9737	464.17	0.2755	0.3819	The accumulated utilities of all of the countries increase, feasible

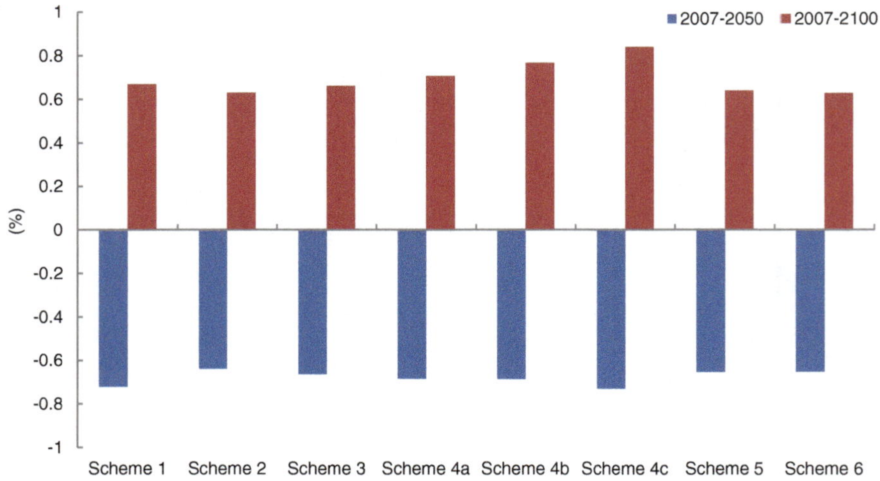

Fig. 6.9 Changing rates of global accumulated utility in schemes (%)

6.5 Conclusions

This study has built a climate change integrated assessment model with global economic interactions: CIECIA. The economic core of CIECIA is a multi-country-sector general equilibrium model, and the climate model is based on the Svirezhev carbon-cycle model. Endogenous technological progress and innovation modes in the field of evolutional economics are introduced into CIECIA to establish the endogenous mode of technological progress, in which the technological progress in the production process is driven by the accumulation of knowledge capital. CIECIA overcomes the defects of ignoring global economic interaction in traditional IAMs, ensuring that the global carbon abatements are discussed and assessed under the condition of global economic general equilibrium. Based on this model, six types of the main global cooperating climate mitigation schemes are simulated, assessed and analyzed. The conclusions derived from the scenario simulations are listed as follows.

1. All of the six types of schemes can achieve the climate mitigation targets by 2100. Thus, they are all effective schemes. There exist differences in the mitigation level of climate change among schemes. Except for Scheme 1, the other schemes are relatively loose, and their global temperature increases by 2100 exceed 1.9 °C.
2. The economic growth of countries are influenced by the implementations of carbon abatements in the early steps. However, after 2050, the economies experience rapid growths because of the lower impact of climate change on economic development. This is the climate welfare brought by carbon abatement.

3. Among these schemes, Scheme 1 benefits the economic developments of the developed countries more, and is unfair to the developing countries for its strict emission reductions. Scheme 2 benefits the economies of developing countries more, especially the middle and low developing countries. However, it has negative influence on the economies of the developed countries, and is unable to benefit all of the participating countries.
4. Scheme 3 benefits countries with huge population bases, e.g., the middle and low developing countries, and fast technological progress. However, it hurts the economies of countries with higher initial carbon emissions per capita, e.g., China and Russia. Although Scheme 4 is the fairest one, it damages the interests of the main economies, e.g., China, the USA, and Russia, etc. and cannot be accepted by all of the abatement participators. Thus, neither of these two schemes are feasible because of the great impacts on the developments of the main economies in the world.
5. Schemes 5 and 6 balance both fairness and feasibility, and have relatively equal impacts on the economies among countries. Scheme 6 realizes the Pareto improvement of the economies of countries in the process of carbon abatement, improving the accumulated utilities between 2007 and 2100 of all of the countries. Despite the sacrificing of some fairness, Scheme 6 can be accepted by all of the reduction participating countries. Thus, Scheme 6 is the most reasonable global cooperating climate mitigation scheme.

Most of the values of the parameters in this study are static except the technological shocks because of the limited space. It is worth noting that many parameter values are controversial, including discount rates, time preferences, and depreciation rates. Even some components of IAM are controversial. For example, there exist different types of the climate damage functions, e.g., the Nordhaus damage function and Weitzman damage function (Weitzman 2009). In fact, we have conducted sensitivity analyses of the parameters and the damage functions. This section will be provided in our next paper. In addition, CIECIA is still a policy evaluation IAM because of solving difficulties, and the commodity trading mechanism is relatively simple for model simplification. The focus of our further study is to realize the optimization calculation and to depict the international commodity flow accurately.

This work was supported by the State Key Development Program on Basic Research of China, No. 2012CB955800

Appendix A. Main Parameters

See Appendix Tables A.1, A.2, A.3 and A.4.

Table A.1 Initial output elasticities of knowledge capital, investing rates and initial knowledge capital stocks

	CHN	USA	JPN	EU	IND	RUS	ODC	HDC	MDC	LDC
$\beta^{Z}_{j,2007}$	0.015	0.043	0.045	0.031	0.01	0.016	0.031	0.016	0.01	0.009
η^{j}	0.0141	0.0257	0.0332	0.0165	0.005	0.01	0.018	0.0014	0.0041	0.0016
Z^{j}_{2007}	162.34	1727.99	696.41	1240.03	24.3	14.03	1569.56	132.74	58.38	13.57

Data source Liu (2013)

Table A.2 Production damage coefficients and global warming damage coefficients

	CHN	USA	JPN	EU	IND	RUS	ODC	HDC	MDC	LDC
$D_{0,j}$	0.1371	0.0992	0.1057	0.1057	0.1371	0.7713	0.1057	0.1144	0.1371	0.115
$b_{1,j}$	0.1	0.07	0.05	0.05	0.1	0.1	0.05	0.1	0.1	0.1

Data source Liu (2013)

Table A.3 Carbon intensities of coal, petro and gas ($MtCO_2$/Mtoe)

	CHN	USA	JPN	EU	IND	RUS	ODC	HDC	MDC	LDC
$\tau^C_{j,coal}$	3.75	3.92	3.71	4.10	3.75	3.53	3.78	3.88	3.77	3.70
$\tau^C_{j,petro}$	2.47	2.69	2.30	2.61	2.44	2.33	2.36	2.69	2.80	2.85
$\tau^C_{j,gas}$	2.21	2.10	2.17	2.09	2.05	2.27	2.14	2.29	2.23	3.28

Data source GTAP-07

Table A.4 Other main parameters

Parameter	Value	Description	Source
ϕ	0.04	Output elasticity of investment	Jin (2012)
θ	1	Constant substitution elasticity in compound functions of consumption, investment and price	Jin (2012)
ρ	0.198	Time preference	Wang et al. (2006)
β	0.015	Discounting rate	Buchner and Carraro (2005) and Wang et al. (2012)
QL_{t0}	1.5202	Carbon emissions from land use in 2007	Nordhaus and Yang (1996)
γ_{le}	0.02	Yearly decay rate of carbon emissions of land use	Nordhaus and Yang (1996)

Appendix B. Changes of Industrial Structure of Countries

Because of the limited space, the 12 sectors in this study are merged into three traditional industries, i.e. the food processing, energy, minerals, light and heavy manufacturing, chemical industry, and construction are merged into secondary industry, and the trade and business services, transport and communication, insurance and finance services, and other Services are merged into the tertiary industry. Table B.1 presents the industrial structures of countries in 2100 under different schemes. The implementations of the abatement schemes have limited effect on the industrial structures of countries in general. The tertiary industry of

Table B.1 GDP shares of the three industries of countries in 2100(%)

		CHN	USA	JPN	EU	IND	RUS	ODC	HDC	MDC	LDC
Scheme 1	Primary industry	2.1624	0.8038	0.7660	1.0687	4.7724	2.0087	1.1193	2.6511	3.6783	6.7285
	Secondary industry	36.6408	26.4365	24.9768	28.4377	37.9948	38.9250	32.2176	35.8221	37.2563	38.8689
	Tertiary industry	61.1968	72.7598	74.2571	70.4937	57.2328	59.0663	66.6631	61.5269	59.0654	54.4025
Scheme 2	Primary industry	2.1476	0.8081	0.7704	1.0747	4.7673	2.0192	1.1258	2.6626	3.6680	6.7659
	Secondary industry	36.6204	26.4427	24.9860	28.4488	37.9837	38.9259	32.2229	35.8268	37.2409	38.8742
	Tertiary industry	61.2320	72.7492	74.2436	70.4765	57.2490	59.0549	66.6513	61.5106	59.0911	54.3598
Scheme 3	Primary industry	2.1872	0.8094	0.7699	1.0724	4.7184	2.0480	1.1235	2.6549	3.6622	6.6553
	Secondary industry	36.6603	26.4445	24.9817	28.4393	37.9508	38.9441	32.2195	35.8165	37.2344	38.8425
	Tertiary industry	61.1525	72.7461	74.2484	70.4883	57.3308	59.0079	66.6569	61.5286	59.1034	54.5022
Scheme 4a	Primary industry	2.1969	0.8242	0.7778	1.0822	4.6375	2.0773	1.1387	2.6538	3.6310	6.6618
	Secondary industry	36.6494	26.4705	24.9819	28.4394	37.8794	38.9449	32.2286	35.7921	37.1846	38.8231
	Tertiary industry	61.1537	72.7052	74.2403	70.4784	57.4832	58.9778	66.6327	61.5541	59.1844	54.5152
Scheme 4b	Primary industry	2.1955	0.8286	0.7786	1.0854	4.6153	2.0853	1.1415	2.6529	3.6187	6.6703
	Secondary industry	36.6390	26.4736	24.9817	28.4381	37.8652	38.9413	32.2266	35.7819	37.1663	38.8167
	Tertiary industry	61.1655	72.6978	74.2396	70.4765	57.5194	58.9734	66.6319	61.5653	59.2150	54.5129
Scheme 4c	Primary industry	2.1933	0.8292	0.7835	1.0903	4.5935	2.0964	1.1462	2.6570	3.6115	6.6747
	Secondary industry	36.6283	26.4664	24.9863	28.4437	37.8463	38.9418	32.2263	35.7770	37.1594	38.8097
	Tertiary industry	61.1785	72.7045	74.2302	70.4660	57.5602	58.9618	66.6274	61.5660	59.2291	54.5156

(continued)

Table B.1 (continued)

		CHN	USA	JPN	EU	IND	RUS	ODC	HDC	MDC	LDC
Scheme 5	Primary industry	2.1546	0.8069	0.7694	1.0736	4.7730	2.0018	1.1253	2.6621	3.6775	6.7178
	Secondary industry	36.6279	26.4399	24.9830	28.4454	37.9868	38.9161	32.2208	35.8244	37.2465	38.8612
	Tertiary industry	61.2175	72.7532	74.2476	70.4809	57.2402	59.0821	66.6540	61.5135	59.0760	54.4210
Scheme 6	Primary industry	2.1586	0.8039	0.7732	1.0808	4.7849	1.9823	1.1317	2.6286	3.6860	6.7078
	Secondary industry	36.6274	26.4303	24.9904	28.4600	37.9914	38.9005	32.2261	35.7924	37.2517	38.8573
	Tertiary industry	61.2140	72.7658	74.2364	70.4592	57.2238	59.1173	66.6422	61.5790	59.0623	54.4349

China has the largest GDP share by 2100 (61.23%) under the Nordhaus Scheme (Scheme 2) while having the least (61.15%) under Scheme 3. The GDP shares of the primary industry and the secondary industry of China by 2100 are higher under Scheme 2 than those under Scheme 3. The tertiary industry of the USA has the largest GDP share by 2100 (72.77%) under the Pareto improvement scheme while having the least (72.70%) in Scheme 4. India obtains the largest GDP share of the tertiary industry by 2100 (57.56%) under Scheme 4c and has the least share (57.22%) under Scheme 6, which is opposite of the structure change in the USA.

As shown in Table B.1, China obtains a higher GDP share of the tertiary industry in the schemes, thus benefiting its development more, e.g. the Nordhaus Scheme, while in Scheme 4, which benefits the economy of India most, the GDP share of tertiary industry of India is higher than those in other schemes. The USA obtains the lowest GDP share of the tertiary industry in Scheme 4, which damages its economic interests most. Thus, the GDP share is closely associated with the economic development of a country. Under the schemes that benefit developing countries more, the developing countries have rapid growth of the economy especially their tertiary industries, competing with the developed countries and accelerating the industrial transfer of the tertiary industry from the developed countries to developing countries. However, under the schemes, which benefit the developed countries more, the economic interests of the developing countries are damaged and the industrial transfer of the tertiary industry decelerates, causing the developed countries to maintain high GDP shares of their tertiary industries.

Appendix C. A New Pareto Improvement Scheme

According to the *U.S.-China Joint Announcement on Climate Change* announced recently, this study designs a new global carbon abatement scheme in which China can reach to its carbon emission peak in 2030 based on the Pareto Improvement Scheme while achieving the Pareto improvements of the accumulated utilities of all of the participating countries. In this scheme (Scheme 7), the carbon emission peak of China is 2030, and the peak emissions are 2.87GtC. China cuts 40% emissions by 2050 based on the 2005 level. The abatement targets of the other countries are the same as those in Scheme 6. In Scheme 7, the carbon emissions of the USA in 2025 is 1.17GtC, having a 28.5% reduction based on 2005 level. That meets the abatement requirements of the USA in the *U.S.-China Joint Announcement on Climate Change*. The abatement tasks of countries are listed in Table C.1.

In Scheme 7, the global mean surface temperature increases by 1.99 °C in 2100 compared with the pre-industrial level, and the atmospheric concentration equivalent of CO_2 is 463.88 ppmv. Thus, Scheme 7 is effective. Figure C.1 presents the

Table C.1 Reduction targets of countries in Scheme 7

	By 2050	By 2100	Carbon permits 2007–2050 (GtC)	Carbon permits 2007–2100 (GtC)
CHN	Start reduction from 2016; Reach carbon peak in 2030; 40% reduction based on 2005 level by 2050	40% reduction based on 2050 level	98.43	134.01
USA	75% reduction based on 1990 level	40% reduction based on 2050 level	45.92	59.60
JPN	90% reduction based on 1990 level	40% reduction based on 2050 level	8.05	9.04
EU	95% reduction based on 1990 level	40% reduction based on 2050 level	28.96	31.11
IND	Reduce to 2005 level	20% reduction based on 2050 level	19.15	33.61
RUS	50% reduction based on 1990 level	40% reduction based on 2050 level	17.71	28.65
ODC	98% reduction based on 1990 level	70% reduction based on 2050 level	15.90	16.13
HDC	20% reduction based on 2005 level	20% reduction based on 2050 level	49.08	87.72
MDC	10% reduction based on 2005 level	20% reduction based on 2050 level	26.00	46.09
LDC	Start reduction from 2045; maintain 2045's emission level between 2046 and 2050	Maintain 2050's emission level	14.31	69.24

accumulated carbon emissions per capita of countries in Scheme 7, which are quite similar to those in Scheme 6. The accumulated carbon emissions per capita of China by 2100 reaches 104.46tC, a little above those in Scheme 6. The Gini coefficients of the carbon emissions in 2100 and the accumulated emissions from 1990 to 2100 are 0.2610 and 0.3990 respectively, a little higher than those in Scheme 6. However, the distances between the Gini coefficients of these two schemes are small; therefore, the fairness of Scheme 7 is acceptable.

Figure C.2 presents the changing rates of the accumulated utilities of the world and countries. All the accumulated utilities of countries from 2007 to 2100 are

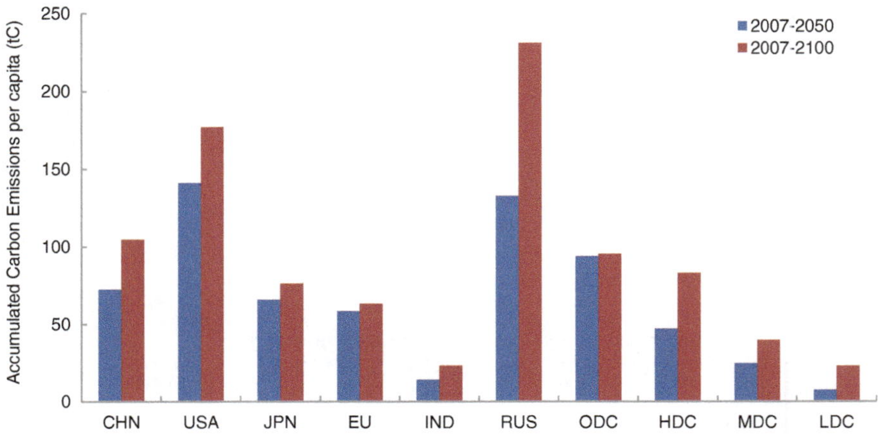

Fig. C.1 Accumulated carbon emissions per capita of countries in Scheme 7 (tC)

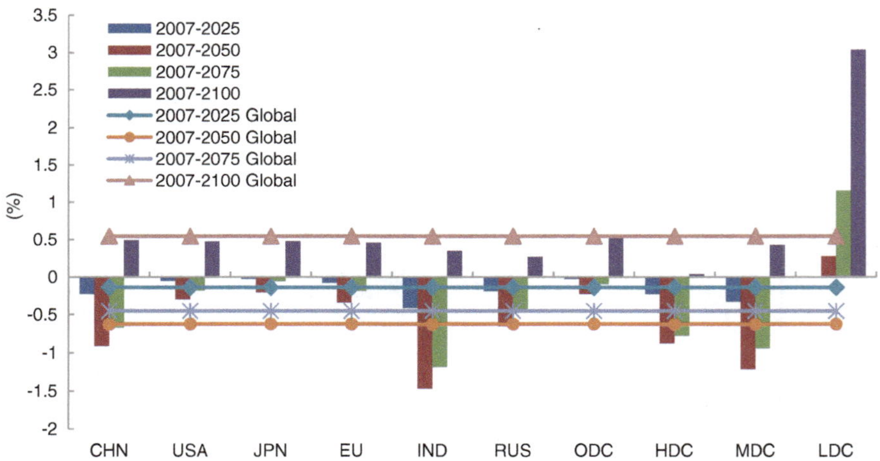

Fig. C.2 Changing rates of accumulated utility in Scheme 7 (%)

higher in Scheme 7 than those in Scheme 0, which means Scheme 7 realizes Pareto improvement, the same as in Scheme 6. The accumulated utility changes of China, the USA and the EU by 2100 are approximately 0.47%, similar to those in Scheme 6; the changing rate of HDC ranks lowest with 0.03%. Because of the more relaxed reduction task, the accumulated utility of China increases obviously compared with that of Scheme 6, avoiding the huge economic losses caused by strict emission controls in the early stages.

Scheme 7 satisfies the feasibility and effectiveness of a global carbon abatement scheme. Although its fairness is lower than that of Scheme 6, the differences between them are very small. Thus, Scheme 7 is reasonable and acceptable to all of the participating countries. This finding proves the abatement rout of the *U.S.-China Joint Announcement on Climate Change* is feasible. In addition, because of the large emissions in the early steps in Scheme 7, China needs to cut 40% of its emissions based on 2005 level from 2030 to 2050, with an emission cut of nearly 2GtC. Thus, China faces heavy abatement tasks in the medium term.

References

Abel AB (2003) The effects of a baby boom on stock prices and capital accumulation in the presence of social security. Econometrica 71(2):551–578

Ackerman F, DeCanio SJ, Howarth RB et al (2009) Limitations of integrated assessment models of climate change. Clim Change 95(3–4):297–315

Akhtar M, Wibe J, Slobodan PS et al (2013) Integrated assessment model of society-biosphere-climate-economy energy system. Environ Modell Softw 49:1–21

Bosetti V, Massetti E, Tavoni M (2007) The WITCH model structure, baseline, solutions. FEEM working paper series No. 10, 2007

Buchner B, Carraro C (2005) Modelling climate policy: perspectives on future negotiations. J Policy Model 227:711–732

Buonanno P, Carraro C, Galeotti M (2003) Endogenous induced technical change and the costs of Kyoto. Resour Energy Econ 25(1):11–34

Chen W, Wu Z, He J (2005) Two-convergence approach for future global carbon permit allocation. J Tsinghua Univ (Sci&Tech), 45(6):850–853

Ding Z, Duan X, Ge Q et al (2009) Control of atmospheric CO_2 concentration by 2050: an allocation on the emission rights of different countries. Sci China Ser D-Earth Sci 39(8):1009–1027

Gu G, Wang Z (2014) A research on the trends of Multi-country-section carbon emissions and energy uses driven by technology progress. Urban Environ Stud 1(1):56–74

He J, Liu B, Chen W (2004) Analysis on the equity of global climate change issues. China Popul Resour Environ 14(6):12–15

IIPCC (2014) IPCC Assessment 5, Mitigation of Climate Change. http://www.ipcc.ch/report/ar5/index.shtml

Jin K (2012) Industrial structure and capital flows. Am Econ Rev 102(5):2111–2146

Kelly DL, Kolstad CD (1999) Integrated assessment models for climate change control. In: Folmer H, Tietenberg T (eds) International yearbook of environmental and resource economics 1999/2000: a survey of current issues. Edward Elgar, pp 171–197

Liu C (2013) The construction of a new style of IAM and study on the global corporations for the mitigation of the carbon dioxide. In: Doctoral degree dissertation. University of Chinese Academy of Sciences, Beijing

Lorentz A, Savona M (2008) Evolutionary micro-dynamics and changes in the economic structure. J Evol Econ 18(34):389–412

Manne AS, Mendelsohn R, Richels R (1995) MERGE: A model for evaluating regional and global effects of GHG reduction policies. Energy Policy 23(1):17–34

Narayanan B, Walmsley TL (2008) Global trade, assistance, and production: the GTAP 7 data base. In: Center for global trade analysis. Purdue University, Washington D.C

Nelson RR, Winter SG (1982) An evolutionary theory of economic change. The Belknap Press of Harvard University Press, Cambridge

Nordhaus WD (1992) An optimal transition path for controlling greenhouse gases. Science 258 (5086):1315–1319

Nordhaus WD (2008) A question of balance: weighing the options on global warming policies. Yale University Press, New Haven

Nordhuas WD, Yang Z (1996) Regional dynamic general-equilibrium model of alternative climate-change strategies. Am Econ Rev 86:741–746

OECD (1997) The OECD GREEN MODEL: AN UPDATED OVERVIEW, Development Centre Working Papers

Paltsev S, Reilly JM, Jacoby HD et al (2005) The MIT emissions prediction and policy analysis (EPPA) model: version 4. MIT joint program for the science and policy of global change report No. 125, Cambridge, MA

Pizer WA (1999) The optimal choice of climate change policy in the presence of uncertainty. Resour Energy Econ 21:255–287

Rotmans J (1990) IMAGE: an integrated model to assess the greenhouse effect. Kluwer, The Netherlands

Schneider SH (1997) Integrated assessment modeling of global climate change: Transparent rational tool for policy making or opaque screen hiding value-laden assumptions? Environ Model Assess 2(4):229–249

Stanton EA, Ackerman F, Kartha S (2009) Inside the integrated assessment models: Four issues in climate economics. Clim Dev 1:166–184

Stem N (2006) Stem review on the economics of climate change. In: Report to the prime minister and the chancellor of the exchequer on the economics of climate change. London, United Kingdom

Stern N (2007) The economics of climate change: the stern review. Cambridge University Press, Cambridge

Svirezhev Y, Brovkin V, Bloh W et al (1999) Optimisation of reduction of global CO_2 emission based on a simple model of the carbon cycle. Environ Model Assess 4:23–33

Tol RSJ (1997) On the optimal control of carbon dioxide emissions: An application of FUND. Environ Model Assess 2:151–163

Tol RSJ (2001) Estimates of the damage costs of climate change. Part 1: benchmark estimates. Environ Resour Econ. 21:47–73

UNDP (2007) Fighting Climate change: human solidarity in a divided world. Human development report, 2007/2008. Oxford University Press

United Nations Department of Economic and Social Affairs (2011) World population prospects: the 2010 revision

Wang Z, Liu X, Tian Y et al (2014) Several issues of climate change ethics. Sci China Earth Sci 44:1600–1608

Wang Z, Zhang S, Wu J (2012) A new RICEs model with the global emission reduction schemes. Chin Sci Bull 57:4373–4380

Wang Z, Jiang Y, Wu J et al (2006) The research on China's potential abatement of CO_2 by technological progress. Acta Ecol Sin 26(2):423–431

Weitzman ML (2009) On modeling and interpreting the economics of catastrophic climate change. Rev Econ Stat 91(1):1–19

Weyant J, Davidson O, Dowlabathi H et al (1996) Integrated assessment of climate change: an overview and comparison of approaches and results. Cambridge University Press, Cambridge

Whitman S, Good G, Donoghue E et al (1997) Mortality in Chicago attributed to the July 1995 heat wave. Am J Public Health 87:1515–1518

Wu J, Zhu Q, Liu C et al (2014) Comparison of the carbon cycle models in DICE/RICE. Acta Ecol Sin 34(22):6734–6744

Zhu Q (2012) Integrated assessment model for climate change with carbon trading mechanism—an agent-Based simulation system. In: Doctoral degree dissertation. University of Chinese Academy of Sciences, Beijing

Zhang S (2012) Design and Implementation of the System for Climate Protection Based on RICE Models of IAM. In: Master's thesis. Graduate University of Chinese Academy of Sciences, Beijing

Chapter 7
Carbon Emission Governance Under Global Carbon Taxes

7.1 Introduction

One of the most essential environment problem human beings need to face is the global temperature rise caused by the emissions of greenhouse gases including CO_2, CH_4 and N_2O from human activities, among which CO_2 is the major one (IPCC 2007). Carbon tax policy has been generally considered as one of the most market effective measures for carbon emission abatement (Baranzini et al. 2000). Compared with other reduction measures, carbon tax brings the *double welfares* of both government revenue and environment. That is, both improving the environmental quality and correcting the tax distortion (Pearce 1991). Besides, the implementation of carbon tax policy can also promote the decline of production cost of firms, stimulate the adoption of energy-saving technologies, and the policy can be adjusted in time if necessary (Pearce 1991). Compared with reduction measures such as carbon emission cap, emission permit trade and emission subsidy, carbon tax policy can run in low quality systems (Brandt and Svendsen 2014), and its effects are better than those of emission cap or permit trade (Avi-Yonahand Uhlmann 2009). Therefore, as a policy mode of climate governance, carbon tax has obtained and long-term and wide supports (Zhang and Baranzini 2004).

The present researches of the carbon tax policies fall into two basic types by their spatial scale: regional and global. The regional carbon tax policies have been studied and discussed widely in academia, including the United Kingdom (Barker et al. 1993), Sweden (Brännlundand Nordström 2004), Ireland (Callan et al. 2009), China (Wang et al. 2009; Yao and Liu 2010) etc. Among them, computational general model (CGE) has become the most important tool in the research on regional carbon tax (Zhang 1998; Wissema and Dellink 2007; Devrajan et al. 2011; Siriwardana et al. 2011).

In the global carbon tax researches, Nordhaus (1993) had a research on the global carbon tax policy by using the DICE model. However, DICE model considers the world as a whole and cannot study the impacts of carbon tax policy at a

Z. Wang et al., *Integrated Assessment Models of Climate Change Economics*, DOI 10.1007/978-981-10-3945-4_7

national or sectoral level. Besides, the endogenous technological mechanism is lacked in DICE. Jaeger (1995) had a research on the changes of global net social welfare cost under the impacts of carbon tax and climate mitigation. As the same as DICE mode, this study considered the world as a whole. Farzin and Tahvonen (1996) looked for the optimal carbon tax rout based on the global carbon cycle model. However, this model merely focused on the consuming effect of fossils, ignoring the consideration of economic factors. Manne and Richels (2006) had a research on the global optimal carbon tax rate under the condition of limited radiative forcing by using the MERGE model. However, this study also failed to analyze economic impacts of carbon tax policy one a country level. The above researches cannot judge the benefit and loss of countries under carbon tax policy, and thus their abatement schemes lack security of fairness in ethics. Silverstein (2010) designed a coordinating carbon tax policy and built a global climate fund by carbon tax revenue that is distributed to countries according to demand factors. However, this study focused on the levy and distribution processes of carbon tax, ignoring the economic impacts of the carbon tax.

Actually, global carbon tax is a global governance problem related to multinational economic development and climate change. Therefore, comprehensive consideration about the global climate change, world economic development and economic interactions among countries is necessary. The international economic relations should not be neglected global carbon tax governance, otherwise, the economic losses of countries suffering under different carbon tax policies cannot be expressed exactly. On the other hand, the economic impact of carbon tax always gets great attentions in the process of carbon tax policy making; and the concern about the economic loss is the fundamental cause of that countries cannot reach an agreement on the carbon tax problem at present. Besides, it is a pity that in the previous researches, the distribution and use of the carbon tax revenue, especially the feasibility of investing carbon tax revenue to low-carbon technological progress, has not drawn enough attention, whereas the distribution of revenue is related to economic benefit that is a major concern of countries and technological progress is an important measure for reducing the economic loss caused by carbon tax and improving the potentials of carbon abatement of countries.

For overcoming the shortcomings of the previous studies, an improved version of Capital, Industrial Evolution and Climate change Integrated Assessment model (CIECIA for short) is built in this study for the simulation of carbon tax policies to have a comprehensive research on the impacts of carbon tax on climate change and human economic activities. CIECIA is a climate-economic integrated assessment model (IAM) developed by Wang et al. (2015) and mainly comprises economic module, climate module, technological progress module etc. Among them, the core of its economic module is a multi-country-sector general equilibrium model that is modified and extended based on the model of Jin (2012)'s two-country model, to reflect the economic relations among sectors of countries. Meanwhile, a stochastic technological progress mode that is driven by knowledge capital is adopted in CEICIA to depict the dynamic relationship between investment and technological progress. In this study, a carbon tax policy module including four carbon tax revenue distribution modes is integrated into CIECIA. Based on this IAM, the impacts of different carbon tax rates, revenue distribution modes and technological progress strategies on the global and national economic developments, carbon emissions and climate change are studied through scenario simulation, and some feasible global carbon governance measures are suggested.

7.2 Model and Data Sources

Because the limitation of space, in this section we only introduce the parts closely related to carbon tax in the model, including carbon tax policy module, technological progress module and the essential parts of the economic model. The details of CIECIA can be seen in Gu and Wang (2015) and Wang et al. (2016). The basic assumptions of CIECIA include: the commodity and capital flow among sectors of countries are free and there is no trading barrier exists; the production sectors in one countries shares a same wage rate and a same labor technological level in one step; sectors producing same products in different countries have a same fixed capital output elasticity; all the countries have a same discounting rate.

7.2.1 Production Module

In this study, country is denoted by *i*, sector is denoted by *j* and step, i.e. year, is denoted by *t* henceforth. A two-layer-nested function formation comprises the Leontief function and the Cobb-Douglas function is employed to depict the relationships among labor, capital stock, value added and gross output. The sectoral gross output is composed of value added and intermediate inputs, and the value added is formed by labor, capital stock and knowledge capital.

$$X_{i,t}^{j} = \min\left\{\frac{M_{1,i,t}^{j}}{a_{1,j,i,t}}, \ldots, \frac{M_{k,i,t}^{j}}{a_{k,j,i,t}}, \ldots, \frac{M_{I,i,t}^{j}}{a_{I,j,i,t}}, X_{i,j,t}^{*}\right\}, \; k = 1, \ldots, I \tag{7.1}$$

$$X_{i,j,t}^{*} = \Omega_{i,j,t}\left(K_{i,t}^{j}\right)^{\alpha_i}\left(A_t^j L_{i,t}^{j}\right)^{1-\alpha_i}\left(Z_t^j\right)^{\beta_{j,t}^{Z}} \tag{7.2}$$

$$Y_{i,t}^{j} = \sum_{k}^{I} M_{k,i,t}^{j} p_{k,t} + X_{i,t}^{j} p_{i,t} \tag{7.3}$$

where $X_{i,t}^{j} p_{i,t}$ represents the value added of sector i of country j in step t; $X_{i,j,t}^{*}$ represents the initial value added formed by labor and capital; $M_{k,i,t}^{j}$ is the intermediate input k in the production process of sector i of country j; $a_{k,j,i,t}$ is the coefficient of intermediate input; $K_{i,t}^j$ is the capital stock; $L_{i,t}^j$ denotes the labor; A_t^j is the labor technology level; α_i is the output elasticity of fixed capital; Z_t^j denotes the knowledge stock; $\beta_{j,t}^Z$ denotes the output elasticity of knowledge capital; $p_{k,t}$ is the price of good k; $\Omega_{i,j,t}$ is the influence of climate change on the economy. Because of free trade among countries, there is one international price associated with each good i in each step.

According to Abel (2003) and Jin (2012), the fixed capital also updates in Cobb-Douglas form. Whereas the update of the knowledge capital is computed with Perpetual Inventory Method.

$$K_{i,t+1}^j = a(I_{K\,i,t}^j)^{\phi}(K_{i,t}^j)^{1-\phi} \tag{7.4}$$

$$Z_{t+1}^j = (1-\delta_Z)Z_t^j + I_{Zt}^j \tag{7.5}$$

where $I_{K\,i,t}^j$ is the fixed capital investment of sector i in country j; $I_{Z\,t}^j$ is the knowledge capital investment; ϕ denotes the output elasticity of in capital production; δ_Z denotes the depreciation rate of knowledge capital. Then the value of fixed capital produced in step t can be found by differentiating Eq. 7.4 with partial respect to $I_{K\,i,t}^j$.

$$q_{i,t}^j K_{i,t+1}^j = 1/\frac{\partial K_{i,t+1}^j}{\partial I_{K\,i,t}^j} K_{i,t+1}^j = \frac{1}{a\phi}\left(\frac{I_{K\,i,t}^j}{K_{i,t}^j}\right)^{1-\phi} K_{i,t+1}^j = \frac{I_{Ki,t}^j}{\phi} \tag{7.6}$$

where $q_{i,t}^j$ denotes the fixed capital price.

According to Jin (2012), the associated price index and the equilibrium prices of products are:

$$\bar{p}_t = \prod_i^I (p_{i,t})^{\gamma_i} = 1 \tag{7.7}$$

$$p_{i,t} = \frac{\gamma_i}{X_{i,t}^g} X_t^g \tag{7.8}$$

The clear condition of the global value added of sectors can be represented by the aggregation of the final consumptions and the investments:

$$X_{i,t}^g = \sum_j^J X_{i,t}^j = \sum_j^J c_{i,t}^j + \sum_j^J \sum_k^I x_{k,i,t}^j \tag{7.9}$$

According to Wang et al. (2012), the Keynes-Ramsey utility function are adopted to reflect the changing of economic strength, which is the analyzing basis of the economic impacts of carbon tax policies.

$$UA^j(T) = \sum_{t=1}^{T} (\beta + 1)^{-t} Pop_t^j \frac{\left(C_t^j/Pop_t^j\right)^{1-\rho}}{1-\rho} \tag{7.10}$$

where $UA^j(T)$ is the cumulative utility of country j until step T, C_t^j denotes the consumption; Pop_t^j denotes the population; β is the discounting rate of which the value is 0.015 consistent with Wang et al. (2012); ρ is the time preference of consumer.

7.2.2 Knowledge Capital and Process Technological Progress

Technological progress is a not negligible economic reality in the reach of long term development. Process technological progress is the process in which the intermediate input demands in production decline brought by technological progress, i.e. the production process innovations that lower the intermediate cost of unitary product in production. Process technological progress is essential to the decline of carbon emission intensities of countries in the economic developments.

According to Lorentz et al. (2008), the loop logarithmic shocking mechanism is adopted in this model to realize the process technological progresses of sectors. The process of the technological shock is: firstly a number from a Uniform distribution $\varepsilon_{j,k,i,t,n} \in N(0;\rho_t^j)$ is drawn by sector i in country j as its technological shock to affect its intermediate input coefficient of product k.

$$\ln(a'_{k,j,i,t}) = \ln(a^*_{k,j,i,t,n-1}) - \varepsilon_{j,k,i,t,n} \tag{7.11}$$

After one turn of technological shocking, a new set of intermediate input coefficients is generated as $\left(a'_{1,j,i,t,N}, \ldots, a'_{I,j,i,t,N}\right)$. This set will be adopted by the sector if the unitary cost of this new set $\sum_k^J a'_{k,j,i,t} p_{i,t}$ is less than the present unitary cost $\sum_k^J a^*_{k,j,i,t,n-1} p_{i,t}$, otherwise the new set will be abandoned as the last one is maintained.

Then, after N turns, the finally obtained $\left(a^*_{1,j,i,t,N}, \ldots, a^*_{I,j,i,t,N}\right)$ is adopted as the intermediate input coefficients of the next step. The variance of the technological shock ρ_t^j depicts the possibility of technological progress and determines the rate of the process technological progress. According to Wang et al. (2015), it is assumed that the process technological progress is influenced by the ratio of knowledge capital in the total capital stock.

$$\rho_t^j = a^\rho \frac{Z_t^j}{Z_t^j + K_t^j} + b^\rho + c_{j,t}^\rho \tag{7.12}$$

where a^ρ, b^ρ and $c_{j,t}^\rho$ are parameters of the function and listed in Sect. 2.5. In Eq. 7. 9, the possibility of technological progress rises by the increase of the share of knowledge capital stock, so that the endogenous mode of technological progress is built. This mode reflects the evolution economic mode that Nelson and Winter (1982) recognized, that is, the technological progress is the result of enterprise's deterministic selection for profit from the microcosmic point, rather than just a slowly incremental process.

Besides, it is worth noting that knowledge capital refers to the macro idea of the accumulation of knowledge, technology and innovation. The improvement of knowledge capital will also raise the production according to Eq. 7.2. Therefore, investing knowledge capital itself is a type of technological progress promotion.

7.2.3 Carbon Emission Accounting and Carbon Tax levy

The energy uses of sectors are all supplied by the energy sector, thus the total energy use in one step can be obtained by aggregating the intermediate energy product inputs of sectors in countries.

$$E_{i,t}^j = \tau_{i,j}^E M_{E,i,t}^j \tag{7.13}$$

where $E_{i,t}^j$ denotes the sectoral energy use; $\tau_{i,j}^E$ is the ratio of energy usage to unitary energy product, i.e. energy intensity. The carbon emission of country j QP_t^j can be obtained by its energy consumption, energy structure and the carbon emission intensities of fossil energies.

$$QP_t^j = \sum_e^E \tau_{j,e}^C \kappa_{e,i,t} \left(E_{C,t}^j + \sum_i^I E_{i,t}^j \right) \tag{7.14}$$

where $\kappa_{e,i,t}$ is the consuming share of the energy that is supplied by fossil energy e of country j in step t; $\tau_{j,e}^C$ is the carbon emission intensity of energy e in country j, $E_{C,t}^j$ is the private energy consumption. The fossil energy is divided into oil, coal and natural gas. The energy structures of countries are obtained by fitting the history data from EIA.

There are two main carbon tax levy ways: on the production side and on the consumption side. Levying carbon tax on the production side is easier to achieve and conducive to the governance of carbon tax and the source control of carbon emission, and is widely accepted in academia. However, in this study we found that the carbon emissions of unitary energy products are different while they are consumed by different sectors in different countries (Table 7.1). Therefore, there exists unfairness in levying carbon tax on the production side because this way of levy ignores the emission differences in consumption.

For this reason, the consumption-side levying way of carbon tax is designed considering the carbon emissions of sectoral productions and private consumptions as the tax base. The specific duty method is adopted and the levy of carbon tax of sector can be given by

$$CT_{i,t}^{j} = \vartheta \sum_{e}^{E} E_{i,t}^{j} \tau_{j,e}^{c} \kappa_{e,i,t} \tag{7.15}$$

where ϑ the tax per unit of carbon emission. According to Hoel (1996), the carbon tax rate across sectors of countries should be unique while the commodities flow freely among countries. Thus, the total carbon tax levy of the world can be given by

$$CT_{t}^{g} = \vartheta \left(\sum_{i}^{I} \sum_{j}^{J} \sum_{e}^{E} \tau_{j,e}^{c} \kappa_{e,i,t} E_{i,t}^{j} + \sum_{j}^{J} \sum_{e}^{E} \tau_{j,e}^{C} \kappa_{e,i,t} E_{C,t}^{j} \right) \tag{7.16}$$

Affected by the carbon tax, the fixed capital margin product changes into

$$dK_{i,t}^{j} = \frac{\alpha_i X_{i,t}^{j} p_{i,t} - CT_{i,t}^{j}}{K_{i,t}^{j}} \tag{7.17}$$

In CIECIA, an overlapping generation model is adopted to depict the consuming and saving behaviors of residents. Therefore, the capital outputs of sectors are all consumed as the retirement pay of the retirees. In this study, we try to shift the carbon tax to the investments the sectors gain for the capital stocks. That is, levying carbon tax from the investments of sectors. Because of the capital price in Eq. 7.6, the sectoral carbon tax levy is

$$I_{i,j,t}^{c-tax-out} = \phi CT_{i,t}^{j} \tag{7.18}$$

where $I_{i,j,t}^{c-tax-out}$ is the tax loss of the investment of sector i in country j.

Carbon tax policy will affect the capital return rates of sectors and thus have significant influence on the international capital flow. The investment mode of CIECIA is composed of investment return rate equilibrium mode and the capital attractiveness mode. Under the influence of carbon tax, the function is changed into:

Table 7.1 Comparisons of sectoral carbon intensities of fossil energies in China and the USA in 2007 ($MtCO_2$/Mtoe)

Fossil energy	China				USA			
	Oil industry	Electricity industry	Chemical industry	Heavy industry	Oil industry	Electricity industry	Chemical industry	Heavy industry
Coal	0.0950	3.8205	3.8060	3.8143	/	3.8775	3.8768	3.8806
Oil	/	3.0458	0.6339	5.3736	/	3.0000	0.4737	/
Natural gas	2.1112	2.2341	1.2300	2.2334	0.4463	2.2274	1.7724	2.2313
Petro	0.3149	2.8734	0.8082	2.6689	1.4424	1.3171	0.2581	2.8933

Data source GTAP-2007

"/" means the consumption or carbon emission of this type of fossil is 0

$$R_1(i,j,t) = \frac{\alpha_i \gamma_i}{s_k - \omega_t^g + (1-\phi)s_l\tilde{s}_{t+1}} \mathrm{E}\left[\frac{X_{i,t+1}^j}{X_{i,t+1}^g}\right] - \frac{\omega_{i,t}^j}{s_k - \omega_t^g + (1-\phi)s_l\tilde{s}_{t+1}} \mathrm{E}\left[\frac{X_{i,t+1}^j}{X_{t+1}^g}\right] + \frac{(1-\phi)s_l\tilde{s}_{t+1}}{s_k - \omega_t^g + (1-\phi)s_l\tilde{s}_{t+1}} \mathrm{E}[R_1(i,j,t+1)] \tag{7.19}$$

$$TK_{i,j}^{x,y} = K_{i,t}^j w_t^y L_t^y \frac{\alpha_x X_{x,t}^y p_{i,t} - CT_{x,t}^y}{K_{x,t}^y} \exp\left(-\upsilon \left| \ln \frac{Y_t^j}{Y_t^y} \right| + 1\right) \tag{7.20}$$

In Eq. 7.19, $R_1(i,j,t)$ is the tax revenue distribution weight under the investment return rate equilibrium mode; $\omega_{i,t}^j$ denotes the ratio of carbon tax levy to the value added of this sector; ω_t^g denotes the ratio of total carbon tax levy to the global value added. In Eq. 7.20, $TK_{i,j}^{x,y}$ is the capital attractiveness intensity from sector x in country y to sector i in country j.

7.2.4 Carbon Tax Revenue Distribution

For satisfying the equilibrium conditions, the carbon tax levy needs to be returned back to sectoral investments or residential consumptions. As CIECIA is a multi-country-sector model, there are two levels in the carbon tax revenue distribution: national and sectoral. In the national level, a global fun pool is set for the national carbon tax revenue distribution uniformly; in the sectoral level, the revenue the countries gain is distributed to sectoral fixed capital investments, consumptions or knowledge capital investments.

In the national level of revenue distribution, we referenced the thoughts of some important global carbon-permit distribution. Four carbon tax revenue distribution modes are designed including the sovereignty principle, the equality principle, the carbon emission per capita principle and the payment ability principle. Meanwhile, because of the specificity of carbon tax, that is different from carbon permit, the carbon tax levy has clear source, we suggest that the tax levies of countries should be considered in the distribution modes.

The sovereignty principle refers to that the revenue is distributed by the carbon tax levies of countries. Then the carbon tax revenue of country j CR_t^j can be given by

$$CR_{j,t}^A = \frac{\xi_t^j / J}{\sum_j^J \xi_t^j / J} CT_t^g = \xi_t^j CT_t^g, \quad \xi_t^j = \frac{CT_t^j}{CT_t^g} \tag{7.21}$$

where ξ_t^j is the share of the tax levy of country j in the global carbon tax revenue; J is the number of country. In Eq. 7.21, the carbon tax levies of countries are directly returned to themselves, thus there is no carbon tax international flow under this principle.

The equality principle refers to distributing carbon tax revenue by the populations of countries in each step.

$$CR_{j,t}^{B} = \frac{\xi_t^j Pop_t^j}{\sum_j^J \xi_t^j Pop_t^j} CT_t^g \tag{7.22}$$

The carbon emission per capita principle refers to distributing carbon tax revenue by carbon emissions per capita of countries. The higher carbon emission per capita the country has, the less revenue its gains.

$$CR_{j,t}^{C} = \frac{\xi_t^j Pop_t^j / QP_t^j}{\sum_j^J \xi_t^j Pop_t^j / QP_t^j} CT_t^g \tag{7.23}$$

The payment ability principle refers to distributing carbon tax revenue by the available resources that the countries can pay. The payment ability is defined as a carbon tax revenue distribution index that is in proportion to the population and in inverse proportion to the value added per capita. Thus, the carbon tax revenue flows from countries with stronger payment abilities to countries whose payment abilities are weak. This principle considers both the population factors and the economic development levels of countries.

$$CR_{j,t}^{D} = \frac{\xi_t^j Pop_t^j \left(X_t^j / Pop_t^j\right)^{-\alpha_e}}{\sum_j^J \xi_t^j Pop_t^j \left(X_t^j / Pop_t^j\right)^{-\alpha_e}} CT_t^g \tag{7.24}$$

where α_e is the parameter of payment ability. According to Wu et al. (2010), the value of α_e is 0.5 in this study.

In the sectoral level, the carbon tax revenue can be returned back to sectoral fixed capital investment $CR_{i,j,t}^{I}$, final consumption $CR_{j,t}^{C}$ and knowledge capital investment $CR_{j,t}^{Z}$.

$$CR_{j,t} = CR_{j,t}^{C} + CR_{j,t}^{I} + CR_{j,t}^{Z} \tag{7.25}$$

We apply a method that distributing carbon tax revenue to sectors according to the carbon emission intensities of sectors, i.e. the higher carbon emission intensity the sector has, the less revenue it gains.

$$CR_{i,j,t}^{I} = \frac{X_{i,t}^j \Big/ QP_{i,t}^j}{\sum_i^I X_{i,t}^j \Big/ QP_{i,t}^j} CR_{j,t}^{I} \tag{7.26}$$

7.2.5 *Data Sources*

The main economic data, including initial value added, fixed capitals, intermediate inputs, and energy consumptions are mainly obtained from GTAP-07 (Narayanan and Walmsley 2008). Capital output elasticity is calculated according to the value added and wages of labor from GTAP-07. The history data of carbon emissions and the carbon intensities of different types of energy of countries are obtained from the website of EIA. Populations of countries and their growth rates are obtained from *World Population Prospects: The 2010 Revision* (United Nations 2011), and the population structures are obtained from the World Bank.[1] The values of knowledge capital related parameters including investing rates and the initial knowledge capital stocks of countries in baseline are cited from Wang et al. (2012) and Liu (2013) and are listed in Table 7.2.

The parameter values of the climate module, including damage function and carbon-cycle model are obtained from to Nordhaus, Yang (1996), Pizer (1999) and Svirezhev et al. (1999).

The values of a^{ρ} and b^{ρ} in Eq. 7.25 are 5.2298×10^{-3} and 1.0499×10^{-4} respectively. Table 7.2 presents the values of $c^{\rho}_{j,t}$. In the regression analysis of Eq. 7.25, R^2 is 0.8740, and the t statistics of the two parameters are −2.89 and 7.75 respectively. The hypothesis of Eq. 7.25 passes the test (Table 7.3).

For simplification, 57 sectors of GTAP-07 are merged into 12, comprising Agriculture (Agri for short), Food Processing (FdPro), Energy (Enrg), Metal and other Minerals (Mtl&Mn), Light Manufacturing (LghtMnfc), Chemical industry (ChemInd), Heavy Manufacturing (HvyMnfc), Construction (Const), Trade and Business Services (Trd&Busi), Transport and Communication (Trans&Comm), Insurance and Finance Services (Ins&Fin) and Other Services (OthServ). The world is also divided into 10 countries/regions, comprising China (CHN, for short), the United States (USA), Japan (JPN), the European Union (EU), Russia (RUS), India (IND), Other developed countries (ODC), High developing countries (HDC), Middle developing countries (MDC), and Low developing countries (LDC).

7.3 Simulations of Different Carbon Tax Rates

According to the thoughts of global governance, the comparison of economic developments and carbon emissions of countries under different carbon tax rates is necessary. In this section, three different global carbon tax rate scenarios, called Scenario Series A (SSA for short), are setting. The tax rates of these scenarios are 10, 20 and 50 US dollars per ton carbon in Scenario A1, A2 and A3 respectively. In SSA, the sovereignty principles are adopted as the tax revenue distribution mode.

[1]Population ages 15–64 (% of total): http://data.worldbank.org.cn/indicator/SP.POP.1564.TO.ZS.

Table 7.2 Related parameter values of knowledge capital

	Output elasticity	Investing rate	Initial stock (billion US Dollar)
CHN	0.015	0.0141	162.34
USA	0.043	0.0257	1727.99
JPN	0.045	0.0332	696.41
EU	0.031	0.0165	1240.03
IND	0.01	0.005	24.30
RUS	0.016	0.01	14.03
ODC	0.031	0.018	1569.56
HDC	0.016	0.0014	132.74
MDC	0.01	0.0041	58.38
LDC	0.009	0.0016	13.57

Table 7.3 Parameter of the relationship between tech-shock and knowledge capital (1×10^{-5})

	CHN	USA	JPN	EU	IND	RUS	ODC	HDC	MDC	LDC
$c_{j,t}^{\rho}$	2.7486	3.1119	−1.8432	2.0534	0.4239	2.4382	−0.4171	−5.5618	−0.5761	−2.3780

Besides, the global carbon tax policies are implemented from 2016. As space is limited, the simulation outcomes of the baseline will not be introduced here.

Figure 7.1 shows the shares of carbon tax in global GDP. It can be seen that the carbon emission intensities decline by process technological progress, leading to the decreasing of the proportions of carbon tax in global GDP. The decreasing rates of carbon tax proportion also decrease generally over time. In Scenario A3, the proportions of carbon tax in global GDP declines from 0.68% in 2016 to 0.38% in

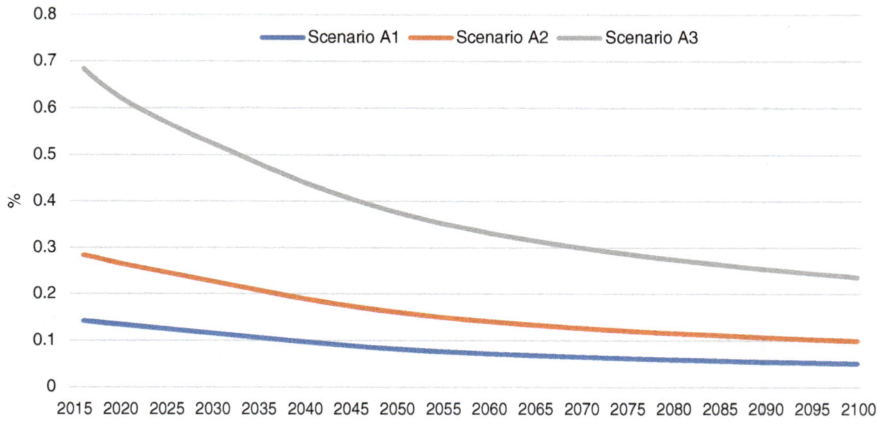

Fig. 7.1 Shares of carbon taxes in global GDP (%)

Table 7.4 Global surface temperature rises from pre-industrial level to Scenario Series A (°)

	2025	2050	2075	2100
Scenario A1	1.1663	1.8067	2.4843	3.1308
Scenario A2	1.1657	1.7956	2.4595	3.0957
Scenario A3	1.1623	1.7526	2.3761	2.9851

2050, and then to 0.24% in 2100, with higher decline rates than those in Scenario A1 and A2.

Table 7.4 shows the global surface temperature rises compared with pre-industrial level in SSA. Along with the increasing of carbon tax rate, the global temperature risings decline. In Scenario A1, the global surface temperature is 3.13 °C higher than pre-industrial level, while in Scenario A3 with a higher carbon tax rate, the temperature rise declines to 2.99 °C, by 0.18 °C from the baseline scenario.

Figure 7.2 shows the changing rates of the product prices of sectors in 2100 from baseline to SSA. Under the influence of carbon tax, the prices of products in highly energy-consumption sectors e.g. Enrg, ChemInd, Mtl&Mn and Trans&Comm rise compared with those in baseline. The rises of prices increase along with the increases of tax rate. Enrg and ChemInd are the most affected sectors by the carbon tax policy. In Scenario 3, the product price of Enrg rises by 3.54%, while the product price of ChemInd rise by 1.45%. However, the rising rates of product prices of Mtl&Mn and Trans&Comm are below 1%.

Figure 7.3 depicts the cumulative carbon reductions of countries from 2016 to 2100 in SSA. Along with increasing of carbon tax rate, the carbon reductions of countries increase obviously. Among them, China, India and HDC have higher reductions than other countries. In Scenario A3, the cumulative carbon reductions of China, India and HDC are 43.42, 36.11 and 16.11GtC respectively, whereas the reductions of Japan, ODC and the USA are only 0.47, 0.60 and 2.39GtC

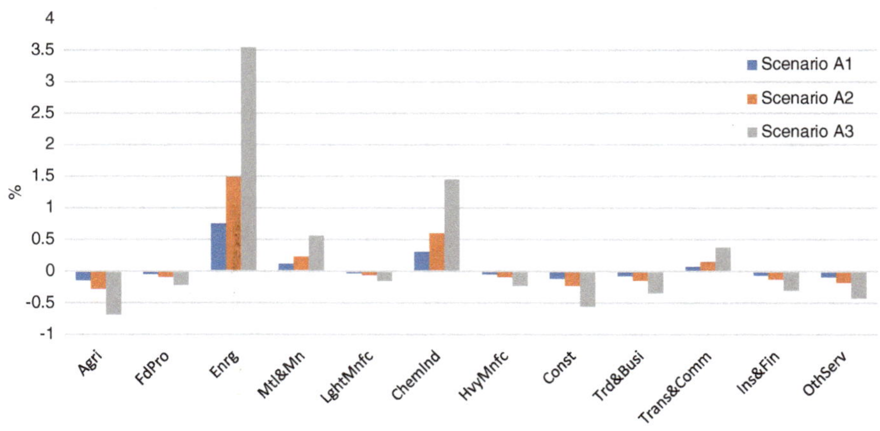

Fig. 7.2 Changes of sectoral product prices in 2100 from baseline to Scenario Series A

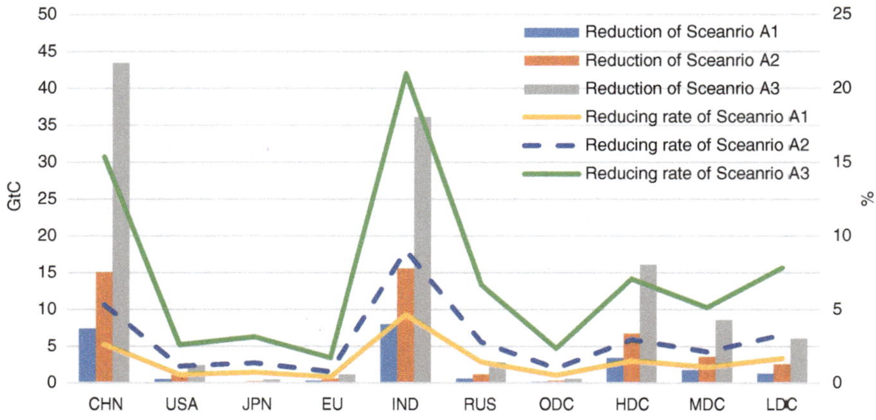

Fig. 7.3 Reductions (GtC) and reducing rates (%) of cumulative carbon emissions between 2016 and 2100 from the baseline to Scenario Series A

respectively. The cumulative carbon reduction rates of China, India and HDC are also much higher than those of developed countries. In Scenario A3, the reducing rates of China and India are 15.35 and 20.99%, whereas those of the USA and the EU are only 2.35 and 1.71%.

There are two main reasons for this phenomenon. First, both China and India are emerging economies. Thus, they have much greater carbon emission demands than developed countries, which brings larger reduction spaces. Second, in developing countries e.g. China and India, the shares of highly energy-consumption sectors in their industrial structures are far above those in developed countries, as well as the carbon emission intensities. Besides, the carbon tax influence on the highly energy-consumption sectors in developing countries makes capital flow tend to flow to developed countries, being a brake of the development of the highly energy-consumption sectors in developing countries somewhat and leading to both higher carbon emission reductions and reducing rates of developing countries further. This phenomenon needs to be given attention in the global governance of carbon abatement.

In Table 7.5, it can be seen that in Scenario A3, the GDP shares of highly energy-consumption sectors e.g. Enrg, Mtl&Mn and ChemInd of China and India in their global sectoral GDPs decline obviously from baseline in 2025. The decline rates of Enrg and Mtl&Mn of China are up to 4.16% and 0.97% respectively, while the decline rate of India's Enrg is 0.83% and the HDC's is 0.30%. Whereas, the GDP shares of highly energy-consumption sectors of developed countries increase generally compared with those in baseline. That means under the influence of carbon tax policies, the capital attractiveness of developing countries for highly energy-consumption sectors weakens, and the trend that the highly energy-consumption sectors shift back to developed countries with advanced low-carbon technologies emerges.

Table 7.5 Changes of GDP shares of high energy-intensive sectors of countries in global sectoral GDPs from the baseline to Scenario A3 (%)

Year	Sector	CHN	USA	JPN	EU	IND	RUS	ODC	HDC	MDC	LDC
2025	Enrg	−4.1548	1.0746	0.2176	1.1396	−0.8229	0.3923	0.3874	1.0064	0.3557	0.4041
	Mtl&Mn	−0.9672	0.3610	0.0821	0.5196	−0.0271	−0.0405	0.1220	0.0308	−0.0370	−0.0437
	ChemInd	−0.3052	0.3421	0.0476	0.5629	−0.0591	−0.1585	0.0551	−0.2989	−0.0935	−0.0926
	Trans&Comm	0.1095	0.0343	0.0721	0.1076	0.0332	−0.0453	0.0185	−0.1712	−0.0953	−0.0633
2050	Enrg	−3.8445	1.6343	0.1180	1.0240	−1.6332	0.5098	0.5820	0.3990	0.5920	0.6186
	Mtl&Mn	−1.1326	0.5557	0.0907	0.6666	−0.0948	−0.0574	0.1755	0.0258	−0.0786	−0.1508
	ChemInd	−0.1209	0.7454	0.0727	0.9330	−0.1502	−0.3307	0.1974	−0.7326	−0.2368	−0.3774
	Trans&Comm	0.4077	0.1041	0.0741	0.1226	0.0811	−0.0711	0.0591	−0.3844	−0.2338	−0.1594
2075	Enrg	−1.2361	1.1872	0.0391	0.5877	−2.0669	0.3831	0.3511	−0.3632	0.5116	0.6064
	Mtl&Mn	−0.4033	0.4351	0.0546	0.4662	−0.1498	−0.0398	0.1221	−0.0718	−0.1290	−0.2844
	ChemInd	0.5009	0.7259	0.0544	0.7784	−0.1254	−0.2516	0.1912	−0.8901	−0.3036	−0.6801
	Trans&Comm	0.5580	0.1285	0.0470	0.0873	0.1131	−0.0499	0.0465	−0.4502	−0.2548	−0.2255
2100	Enrg	0.1170	0.8606	0.0208	0.3894	−2.0593	0.3019	0.2002	−0.8446	0.4486	0.5654
	Mtl&Mn	−0.0077	0.3387	0.0363	0.3401	−0.1237	−0.0179	0.0850	−0.1361	−0.1099	−0.4048
	ChemInd	0.7352	0.6547	0.0450	0.6375	0.0112	−0.1403	0.1597	−0.8934	−0.2609	−0.9488
	Trans&Comm	0.5440	0.1261	0.0296	0.0769	0.1493	−0.0293	0.0313	−0.4561	−0.1976	−0.2743

Along with the weakening of carbon tax's economic impacts and the process technological progress in developing countries, the industrial structure changes of countries brought by the carbon tax policies decrease. In 2075, the drop of the GDP share of China's Enrg from the baseline decreases to 1.24 percentage points, and its GDP share exceeds that of the baseline in 2100. The drop of GDP share of China's Mtl&Mn from the compared with the baseline decreases from 0.97 percentage point in 2025 to 0.01 percentage point in 2100; the change of GDP share of India's ChemInd rises from −0.15 percentage point to 0.01 percentage point in 2100; the GDP share change of HDC's Trans&Comm rise from −0.25 percentage point in 2075 to −0.20 percentage point in 2100.

Due to the limitation of space, this section only analyzes the changing rates of cumulative utilities of countries in Scenario A3 to have a study on the economic impacts of carbon tax policy. From Fig. 7.4, it can be seen that in the early period of the simulation, the cumulative utilities of countries decline at different levels compared with the baseline scenario, which indicates that the carbon tax policy has negative impacts on the economic developments, hurting the economic benefits of countries. This kind of hurt peaks between 2050 and 2075. The loss rates of cumulative utilities of countries decline obviously in 2100 compared with those in 2075. That is mainly because of the climate welfare causing by the carbon emission reduction which leads to the decreasing of global surface temperature rising. Thus, the climate welfare can compensate the economic loss of carbon tax policies to some extent. However, the carbon tax revenue in SSA is all consumed, rather than invested into sectoral productions or to promote technological progress. Therefore, the cumulative utilities of countries between 2007 and 2100 in SSA are still below those in the baseline scenario. That is disadvantageous to economic developments of countries.

It is worth noting that the carbon emission reductions of countries in SSA are much less than those of Pearce (1991) and Jaeger (1995) in which the carbon tax

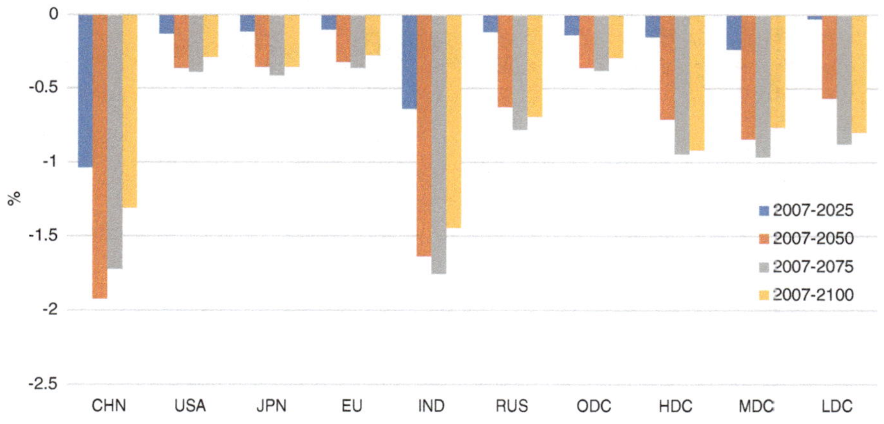

Fig. 7.4 Changes of cumulative utilities of countries between 2007 and specified year from baseline to Scenario A3 (%)

results from several global IAMs researches including GREEN, MERGE and DICE are reviewed, especially in developed countries. That is mainly because the implementation time of carbon tax policies in this study is 2016, quite later than those studies above. In the last 20 years, most of developed countries peaked or had peaked their carbon emissions, and their carbon emissions have shown a trend of decline at present. Thus, the impacts of carbon tax policy today are much weaker than those of more than 20 years ago. Besides, the currency the above studies used is the UD dollars in constant 1990's prices. Considering the inflationary factors, the real carbon tax rate in the above studies are higher than this study.

7.4 Simulations of Different Distribution Modes of Carbon Tax Revenue

Based on the introduction of the distribution modes of carbon tax revenue, this section simulates and analyzes the impacts of the distribution modes of carbon tax revenue on the developments and carbon emissions of countries. Four scenarios following four different distribution principles, called Scenario Series B (SSB for short), are setting in this section: sovereignty principle scenario (Scenario B1), equality principle scenario (Scenario B2), carbon emission per capita principle scenario (Scenario B3) and payment ability principle scenario (Scenario B4). The carbon tax rate in SSB is 20 US dollars per ton carbon, and the policies are still implemented from 2016. Different from SSA, in SSB 50% of the carbon tax revenue of countries is used to consume, while the other 50% is allocated to the sectoral investments of fixed capital stocks following Eq. 7.23.

Table 7.6 shows the global surface temperature rises in SSB. It can be seen that carbon tax revenue distribution mode has very little influence on the global temperature rise. In SSB, the global surface temperatures in 2100 rise around 3.10 °C from the pre-industrial level. Because a part of the tax revenue returns back to the sectoral investments of fixed capital stock, the temperature rises in SSB increase slightly from Scenario A2, but are still lower than Scenario A1.

Figure 7.5 depicts the cumulative carbon revenue flows among countries. There is no international carbon tax revenue flow in Scenario B1 because the tax revenue is distributed to countries according to the carbon tax incomes of countries. In Scenario B2, China, India, MDC and LDC become the net inflow countries of carbon tax revenue countries, whereas the revenue of the USA outflows largely. In

Table 7.6 Global surface temperature rises from pre-industrial level to Scenario Series B (°C)

	2025	2050	2075	2100
Scenario B1	1.1657	1.7964	2.4627	3.1017
Scenario B2	1.1657	1.7964	2.4629	3.1021
Scenario B3	1.1657	1.7963	2.4624	3.1017
Scenario B4	1.1657	1.7964	2.4628	3.1022

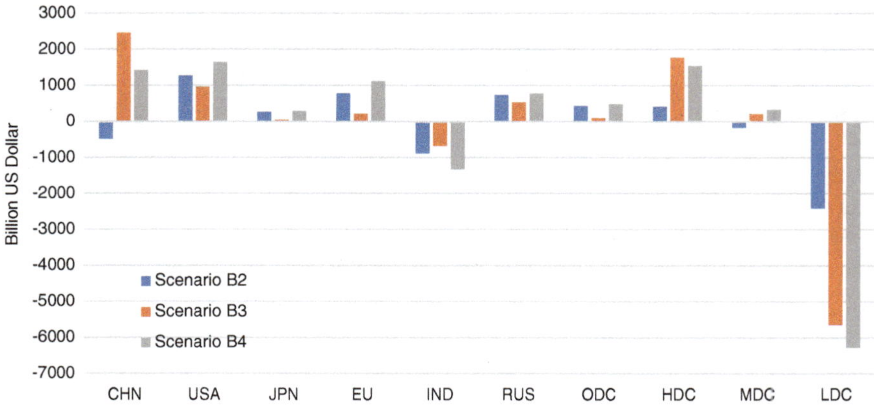

Fig. 7.5 Cumulative net carbon tax outflow of countries from 2016 to 2100 in Scenario Series B (10^8 US. Dollars)

Scenario B3, only India and LDC keep the net inflow trend of carbon tax revenue, whereas China becomes the largest source of the outflow of carbon tax revenue. The net inflow of LDC's carbon tax revenue rises significantly from Scenario B2 to Scenario B3. In Scenario B4, the net outflow of China's tax revenue declines compared with Scenario B3, the USA becomes the largest net outflow country again, and India and LDC gain more revenue net inflow than other scenarios.

Under the principle of equality, China is a net inflow country of carbon tax revenue whose cumulative inflow of carbon tax revenue is higher than its cumulative outflow. Thus, the equality principle can benefit the economic development of China. To developed countries with lower populations and higher GDPs per capital, the payment ability principle is the most negative one. However, countries with huge population sizes and lower economic developing levels e.g. India and LDC will benefit most from this principle. Because of its decreasing population after 2025 and the high level of carbon emission demand under the rapid growth of economy, the economic benefit of China will be hurt most under the carbon emission per capita principle. From the perspective of global governance, the carbon emission per capita principle just pursues the fairness of carbon emission distribution, ignoring the consideration of the economic statuses and developing demands of countries in reality. Under carbon emission per capita principle, the net outflows of carbon tax revenue of developed countries are lower than those under equality principle and payment ability principle because the developed countries are industrialized and have much higher energy technological levels, whereas the net outflows of China, India and HDC are higher because these countries are still during rapid industrialization with increasing carbon emissions per capita in future. Therefore, the carbon emission per capita principle is unfair actually.

Besides, considering the ratios of carbon tax revenue to economic sizes, the Russia's outflows of carbon tax revenue in SSB are quite incredible. That is mainly because the highly energy consumption sectors e.g. Enrg, Mtl&Mn, ChemInd occupy large accounts in its industrial structure.

Figure 7.6 shows the changes of cumulative utilities of countries between 2007 and 2100 from the baseline scenario to SSB. The changes of the cumulative utilities are similar to the changes of the net flows of carbon tax revenue, which indicates the carbon tax policy will affect the economic development of countries directly. In Scenario B1, as a part of the tax revenue is invested back to sectors, the cumulative utilities of developed countries and Russia that are less affected by the carbon tax policies exceed those in baseline by 2100. In Scenario B2, as becoming the carbon tax revenue net inflow countries, the cumulative utilities of India and LDC increase compared with the baseline. The decreasing rate of China's cumulative utility decline to 0.05% from the baseline to Scenario B2. In Scenario B3, China suffers large drop of cumulative utility by 0.25% from the baseline, with is in accord with its state of massive carbon tax revenue outflow, while the cumulative utility of LDC rises by 1.21%. In Scenario 4, developed countries and Russia suffer higher decline rates of cumulative utilities, which is also accord with their states of carbon tax revenue flow.

It is worth noting that contrary to the differences of the cumulative carbon tax revenue inflow, the rise of cumulative utility of LDC in Scenario B4 is lower than that of Scenario B3, while the cumulative utility of China in Scenario B4 is higher than those of Scenario B1. This phenomenon is related to the over-time changes of carbon tax revenue distribution. Although the cumulative carbon tax revenue flow of China is negative in Scenario B4, China is a net inflow country of carbon tax revenue in the early steps before 2035 (Table 7.7). Thus, China gains more tax revenue in Scenario B4 than in Scenario B1. According to Eq. 7.7, because of the existence of discount rate, the longer the future utility is from current, the less current utility it is after discounting. Therefore, despite keeping net outflow state of tax revenue after 2035, China's cumulative utility between 2007 and 2100 in

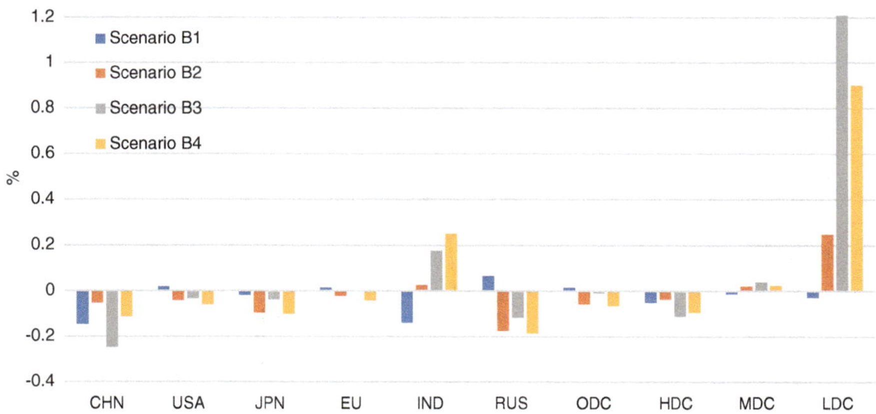

Fig. 7.6 Changes of cumulative utilities of countries between 2007 and 2100 from the baseline to SSB (%)

Table 7.7 Net inflows of carbon tax revenue in China at early steps in Scenario B1 and B4 (10^9 US. Dollars)

Scenario B1					Scenario B4				
2016	2020	2025	2030	2035	2016	2020	2025	2030	2035
0	0	0	0	0	32.4001	25.5558	16.7847	7.6309	−1.0138

Table 7.8 Net inflows of carbon tax revenue in LDC at early steps in Scenario B3 and B4 (10^9 US. Dollars)

Scenario B3					Scenario B4				
2016	2020	2025	2030	2035	2016	2020	2025	2030	2035
34.3214	37.4542	41.8798	46.4959	50.5840	13.8628	16.6606	21.0015	26.3500	32.6626

Scenario B4 is still higher than in Scenario B1 in which no carbon tax revenue flow exists depending on the net inflow of carbon tax revenue before 2035. Similarly, although the total cumulative inflow of carbon tax revenue of LDC in Scenario B4 is higher than in Scenario B3, its cumulative utility between 2007 and 2100 is less instead because of its lower carbon tax revenue inflow in the early period (Table 7.8).

In this section, the outcomes of the carbon emissions in SSB are compared with those in Scenario A2 in analysis because of the same carbon tax rate and similar carbon emissions of countries. Figure 7.7 shows that the cumulative carbon emissions of countries in SSB increase slightly from Scenario A2. In general the carbon emission of China increases most, by more than 1.1GtC, followed by HDC and MDC whose carbon emissions increase by more than 0.7GtC, whereas the rises of carbon emissions of Japan and ODC are negligible. The carbon emissions of developed countries, Russia and LDC increase most under the sovereignty

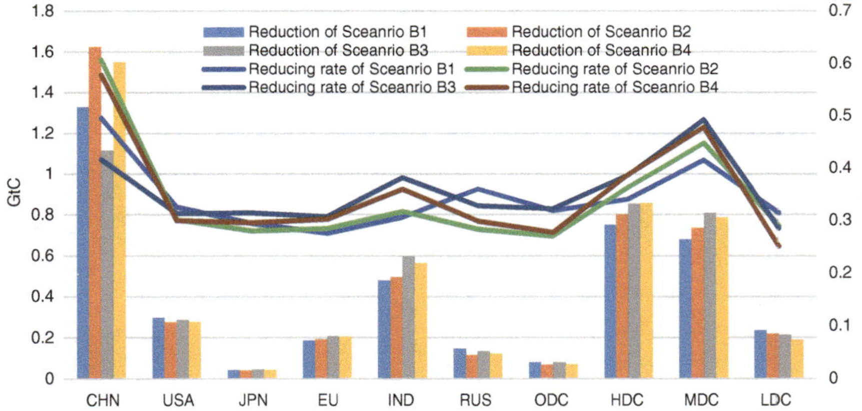

Fig. 7.7 Increases (GtC) and increasing rates (%) of cumulative carbon emissions of countries between 2016 and 2100 from Scenario A2 to SSB

principle, while the carbon emissions of India and MDC increase most under the principle of carbon emission per capita. China raises its carbon emission most under the equality principle.

The changes of carbon emission of India and LDC are more complicated. On one hand, inflow of carbon tax revenue will raise the carbon emissions by more investments to sectors, but on the other hand, more revenue inflow will promote economic growth and then accelerate the technological progress, bringing down the carbon emission intensions. That is why after massive carbon tax revenue inflow in Scenario B2, B3 and B4, the carbon emission of LDC are lower than that of Scenario B1, and in Scenario B3 in which India gains its least revenue inflow except in Scenario B1, India's carbon emission is higher than in Scenario B2 and B4 instead.

7.5 Impacts of Technological Progress Strategy in Carbon Tax Policy

This section has a research on the changes of economic developments and carbon emissions of countries under the condition that a part of the carbon tax revenue is invested to the knowledge capital stocks for improving the rates of technological progress. One important reason of that the global carbon tax policy receives more and more support is the potential benefit brought by the investment in the energy technologies by using carbon tax revenue (Rees 2006; Schlesinger 2006; IPCC 2007). Three scenarios called Scenario Series C (SSC for short) are setting in this section. In Scenario C1, 20% of the carbon tax revenue of countries is invested in their knowledge capital stocks to promote technological progress, while 40% in Scenario C2 and 50% in Scenario C3. In SSC, the carbon tax rate is 20 US dollar per ton carbon; the revenue distribution mode is the equality principle; 50% of the tax revenue of countries is invested to the fixed capital stock of sectors and the others are consumed; the carbon tax policy is still implemented from 2016.

Table 7.9 shows the global temperature rises in SSC. Compared with those in SSA and SSB, the temperature rises in SSC decline significantly under the effect of the acceleration of the process technological progress rates of countries. In Scenario C1, the global temperature rise in 2100 is 3.01 °C, lower than that of Scenario A3 in which the carbon tax rate is up to 50 US dollar per ton carbon. In Scenario C3, the global temperature rise in 2100 decline to 2.87 °C, decreasing by 0.3 °C from the same step in the baseline. The effect of temperature control of SSC is obvious.

Table 7.9 Temperature rises from pre-industrial level in Scenario Series C (°C)

	2025	2050	2075	2100
Scenario C1	1.1657	1.7866	2.4191	3.0090
Scenario C2	1.1657	1.7758	2.3739	2.9160
Scenario C3	1.1657	1.7703	2.3522	2.8726

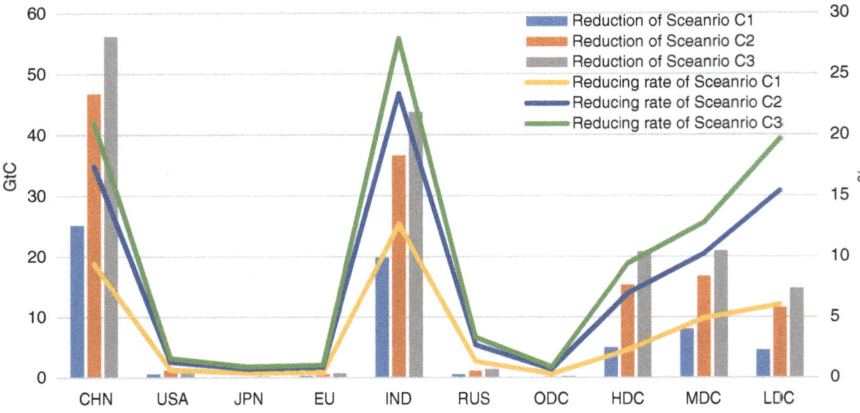

Fig. 7.8 Reductions (GtC) and reduction rates (%) of cumulative carbon emissions of countries from Scenario A2 to Scenario Series C

Figure 7.8 shows the carbon emission reductions and reducing rates of countries from Scenario A2 to SSC. Along with the rising of knowledge capital investment, the reductions and reducing rates increase somewhat. In Scenario C3, the carbon reductions of China and India reach to 56.19 and 43.71GtC respectively compared with Scenario A2, more than other countries, and India, China and LDC have higher reducing rates that are 27.92%, 20.98% and 19.66% respectively. Both the carbon reductions and reducing rates of developed countries and Russia are much lower than those of developing countries. In Scenario C3, the reductions of developed countries and Russia are all less than 1.5GtC, and the reducing rates are all below 4%.

It is indicated that the developing countries, especially those with higher carbon emission demands in future as China and India, are more sensitive to the improvement of knowledge capital and will achieve higher carbon reductions, whereas the reductions of developed countries and Russia are very limited, which means those countries are insensitive to the improvements of knowledge capital investing rates. The main reasons of this phenomenon are: first, the carbon emission demands of developed countries and Russia are lower than the other developing countries, reflecting limited reduction potentials of developed countries and Russia. Second, developed countries always have higher process technological levels as well as knowledge capital stocks at the initial step, which makes the added knowledge capital from the carbon tax revenue have very little effect on the process technological progress rates of developed countries.

Beside, under the distribution principle of equality, developed countries and Russia are all net outflow countries of carbon tax revenue. This also limits those countries, especially Russia, to invest their tax revenue to knowledge capital stocks. It can be seen in Fig. 7.5 that the Russia's cumulative net outflow of carbon tax revenue is as same as that of the EU, much higher than Japan and ODC, although the economic size of Russia is quite smaller than EU. Table 7.10 shows that the

Table 7.10 Ratios of cumulative carbon tax revenue inflows to outflows of countries in SSC

	CHN	USA	JPN	EU	IND	RUS	ODC	HDC	MDC	LDC
Scenario C1	1.1462	0.3259	0.1019	0.4380	1.3370	0.1115	0.1579	0.9401	1.0885	2.6545
Scenario C2	1.1346	0.3239	0.1013	0.4355	1.3247	0.1108	0.1571	0.9318	1.0797	2.6383
Scenario C3	1.1106	0.3201	0.1003	0.4306	1.2995	0.1094	0.1554	0.9154	1.0624	2.6063

ratios of Russia in SSC are close to Japan's levels, lower than the ratios of the USA and the EU and especially lower than developing countries e.g. China, India and LDC. That indicates the carbon tax revenue of Russia is very limited, leading to fewer added knowledge capital from the carbon tax revenue. This makes Russia insensitive to the technological progress policy like developed countries in SSC to some extent, although the initial process technological level and knowledge capital stock of Russia are both far from developed countries.

Figure 7.9 shows the changing rates of cumulative utilities of countries between 2007 and 2100 from the baseline scenario to SSC. Under the influence of technological progress, the cumulative utilities of countries in SSC are higher than any scenario in SSA or SSB. In Scenario C3, almost all the cumulative utilities are higher than the baseline, which means the economic losses by carbon tax can be fully compensated by the impacts of technological progress and the climate welfare. Similar to Fig. 7.8, the sensitivities of developing countries to the policy are much higher than those of developed countries and Russia. In Scenario C3, LDC raises its cumulative utility by 11.32% from the baseline, higher than any other countries. It is followed by HDC and MDC which raise their cumulative utilities by 6.86% and 4.70% respectively. Whereas, the changing rates of the USA and the EU in

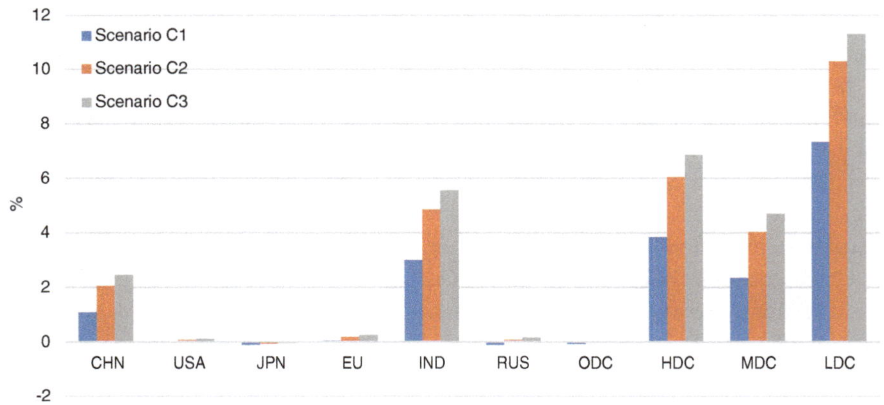

Fig. 7.9 Changes of cumulative utilities of countries between 2007 and 2100 from the baseline to Scenario Series C (%)

Scenario C3 are rarely 0.12 and 0.25%, and the cumulative utility of Japan in Scenario C3 is still below the baseline level.

To developed countries, their higher knowledge capital and process technological levels and lower gained carbon tax revenue are still the main reasons why the impact of the strategy using carbon tax revenue to invest knowledge capital on the knowledge capital stocks and process technological levels is so weak. To Russia, its main reason of weak changes of cumulative utilities in SSC is its extremely low carbon tax revenue.

7.6 Conclusions

In this study, a new improved version of CIECIA combining with a carbon tax policy module was built. In carbon tax module, four distribution modes of carbon tax revenue were designed according to different distribution principles. Based on the scenario simulations, the impacts of global carbon tax policy on the economic developments of sectors in countries and the global climate changes are studied, as well as the impacts of different distribution modes and technological progress strategies.

The results show that the carbon tax policies would obviously promote the carbon emission reductions of countries, especially the developing countries. In Scenario A3 while the carbon tax rate is 50 US dollar per tonne carbon, the reductions of cumulative carbon emissions of China and India between 2016 and 2100 can reach to 43.42 and 36.11GtC, while the carbon reductions of the USA and Japan are 2.39 and 0.47GtC respectively. The main reason of this phenomenon is that China and India have higher carbon reduction potentials for their higher initial carbon emission intensities and greater carbon emission demands in future. Besides, the highly energy-consumption sectors of developing countries will transfer to developed countries in which the economic losses under carbon tax policy are fewer. This enlarges the distance of carbon reduction between developing countries and developed countries to some extent.

In the four international carbon tax revenue distribution modes, the equality principle benefits the carbon tax revenue of China most. The carbon emission per capita principle benefits developed countries and the middle and low developing countries, whereas leading to tax revenue losses in China, India and HDC. That is because the carbon emission per capita principle pursuits the equality in revenue distribution unilaterally, ignoring the development demands of emerging economies as China and India. Because of the decreasing trend of population after 2025 and the continuingly large carbon emission demand along with the rapid economic development, China suffers the greatest loss of carbon tax revenue under the principle of carbon emission per capita. The payment ability principle is the most disadvantageous to the carbon tax revenue of developed counties with small populations but high GDPs per capita. However, it benefits countries with large population sizes and low economic levels as India and LDC. In sum, under the equality principle, the net inflows of carbon tax revenue of countries are closer. Under the payment ability principle, the net outflows of carbon tax revenue of developed countries are higher than under the equality principle and the carbon emission per capita principle, while the inflows of developing countries, especially LDC rise obviously.

Investing knowledge capital stock buy using carbon tax revenue to promote the process technological progress would improve the carbon emission reductions of countries effectively, especially for developing countries. While the share of knowledge capital investment in the carbon tax revenue rise up to 50%, the cumulative reductions of China and India from the baseline reach to 71.19 and 59.23GtC respectively. Developed countries and Russia are almost insensitive to the knowledge capital investment measures and reduce their carbon emissions very few in Scenario Series C because of several reasons e.g. less reduction potentials led by higher initial knowledge capitals and process technological levels, and less carbon tax revenue.

It is worth noting that even when the share of knowledge capital investment is up to 50% in Scenario C3, the global surface temperature rise still reaches to 2.87 °C, far from the 2 °C global warming controlling target of the internationally recognized Copenhagen Consensus. Thus, the carbon tax policy can only reduce the carbon emissions to a certain extent, but not solve the pressing global climate change problems fundamentally. Therefore, the implementation of global production abatements is still an important measure for achieving the goal of global climate mitigation.

References

Abel AB (2003) The effects of a baby boom on stock prices and capital accumulation in the presence of social security. Econometrica 71(2):551–578

Avi-Yonah RS, Uhlmann DM (2009) Combating global climate change: why a carbon tax is a better response to global warming than cap and trade. Stanford Environ Law J 28(1):3–50

Baranzini A, Goldemberg J, Speck S (2000) A future for carbon taxes. Ecol Econ 32(3):395–412

Barkera T, Baylisb S, Madsenb P (1993) A UK carbon/energy tax: the macroeconomics effects. Energy Policy 21(3):296–308

Brandt US, Svendsen GT (2014) A global CO_2 tax for sustainable development? J Sustain Dev 7(1):85–93

Brännlund R, Nordström J (2004) Carbon tax simulations using a household demand model. Eur Econ Rev 48(1):211–233

Callan T, Lyons S, Scott S, Tol RSJ, Verde S (2009) The distributional implications of a carbon tax in Ireland. Energy Policy 37(2):407–412

Devarajan S, Go D, Robinson S, Thierfelder K (2011) Tax policy to reduce carbon emissions in a distorted economy: illustrations from a South Africa CGE model. B.E. J Econ Anal Policy 11(1):1–22

Farzin YH, Tahvonen O (1996) Global carbon cycle and the optimal time path of a carbon Tax. Oxford Econ Papers New Ser 48(4):515–536

Gu G, Wang Z (2015) Research of Global Economy and Climate Protection under Economic Interaction and Industrial Revolution. Science Press, Beijing

Hoel M (1996) Should a carbon tax be differentiated across sectors? J Public Econ 59(1):17–32

Intergovernmental Panel on Climate Change (IPCC) The 4th Assessment Report [R] (2007). http://www.ipcc.ch

Jaeger WK (1995) The welfare cost of a global carbon tax when tax revenues are recycled. Resour Energy Econ 17:47–67

Jin K (2012) Industrial structure and capital flows. Am Econ Rev 102(5):2111–2146

Liu C (2013) The construction of a new style of IAM and study on the global corporations for the mitigation of the carbon dioxide. Doctoral degree dissertation. University of Chinese Academy of Sciences, Beijing

Lorentz A, Savona M (2008) Evolutionary micro-dynamics and changes in the economic structure. J Evol Econ 18(34):389–412

Manne AS, Richels RG (2006) The role of non-CO_2 greenhouse gases and carbon sinks in meeting climate objectives. Energy J (Multi-Greenhouse Gas Mitigation and Climate Policy, special issue no. 3) 3:393–404 (2006)

Narayanan B, Walmsley TL (2008) Global trade, assistance, and production: the GTAP 7 data base. Center for Global Trade Analysis, Purdue University, Washington DC

Nelson RR, Winter SG (1982) An evolutionary theory of economic change. The Belknap Press of Harvard University Press, Cambridge

Nordhaus WD (1993) Association optimal greenhouse-gas reductions and tax policy in the "DICE" model. Am Econ Rev 83(2):313–317

Nordhuas WD, Yang Z (1996) Regional dynamic general-equilibrium model of alternative climate-change strategies. Am Econ Rev 86:741–746

Pearce D (1991) The role of carbon taxes in adjusting to global warming. Econ J 101(407): 938–948

Pizer WA (1999) The optimal choice of climate change policy in the presence of uncertainty. Resour Energy Econ 21:255–287

Rees M (2006) The G8 on energy: too little. Science 313:591

Schlesinger W (2006) Carbon trading. Science 314:1217

Silverstein DN (2010) A method to finance a global climate fund with a harmonized carbon tax. Unpublished. http://mpra.ub.uni-muenchen.de/27121/

Siriwardana M, Meng S, McNeill J (2011) The impact of a carbon tax on the Australian economy: results from a CGE model. Bus Econ Public Policy Working Papers, 2. School of Business, Economics and Public Policy, University of New England

Svirezhev Y, Brovkin V, Bloh W, Schellnhuber HJ, Petschel-Held G (1999) Optimisation of reduction of global CO_2 emission based on a simple model of the carbon cycle. Environ Model Assess 4:23–33

United Nations Department of Economic and Social Affairs (2011) World population prospects: the 2010 revision. http://www.un.org/en/development/desa/publications/index.html

Wang J, Yan G, Jiang K, Liu L, Yang J, Ge C (2009) The study on China's carbon tax policy to mitigate climate change. China Environ Sci 29(1):101–105

Wang Z, Zhang S, Wu J (2012) A new RICEs model with the global emission reduction schemes. Chin Sci Bull 57:4373–4380

Wang Z, Gu G, Wu J, Liu C (2015) CIECIA: a new climate change integrated assessment model and its assessments of global carbon abatement schemes. Sci China: Earth Sci 58:1–22

Wang Z, Gu G, Wu J, Liu C (2016) CIECIA: a new climate change integrated assessment model and its assessments of global carbon abatement schemes. Sci China: Earth Sci 59:185–206

Wissema W, Dellink R (2007) AGE analysis of the impact of a carbon energy tax on the Irish economy. Ecol Econ 61:671–683

Wu J, Ma X, Wang Z (2010) Provincial emission permits quota allocations in China. Quat Sci 30(2):481–488

Yao X, Liu X (2010) Optimal carbon tax in China with the perspective of economic growth. Econ Res J 11:48–58

Zhang Z, Baranzini A (2004) What do we know about carbon taxes? An inquiry into their impacts on competitiveness and distribution of income. Energy Policy 32(4):507–518

Zhang Z (1998) Macro-economic and sectoral effects of carbon taxes: a general equilibrium analysis for China. Econ Syst Res 10(2):135–159

Chapter 8
Global Climate Ethics: A View Based on Chinese Philosophy

The Ethical Factors have been recognized in the scientific researches on the Climate Change. Active responding to the Climate Change is a common goal and mission of all human beings. The better approach of scientific researches on the Climate Change should integrate the studies on economic realities as well as ethics. Focused on such researches on the national and international levels, this essay is trying to introduce a more rational climate ethical system, on the basis of traditional Chinese Ethics, with emphasis on its core value of "harmony" and the practice of "operation on ranking and grouping". Under the practice, this essay probes into the fundamental ethical principles in the climate policy making—Equity, Justice, Values and Wellbeing, pinning the ethical definitions of Equity and Justice to set the ethical criteria in the negotiations on the Climate Change. In the end, it draws to a conclusion of Climate Ethics of the global community: all the measures against the Climate Change should keep in line with the social and economic development of all the participating countries, paving the way for the Pareto improvements in each participant while reducing the gaps among them, without depressing the social individuals which are being poor deeper into poverty.

8.1 Introduction

The Climate Change has come into the spotlight of the global community as it holds an influence across any social or political boundaries, and as well as one of the key factors to the development of every country, even to each person's life. Thus, it demands every government to promulgate its policy/policies on it. Regardless of the objections and denials from the international NGOs, like NIPCC, the World Meteorological Organization (WMO) and the United Nations Environment

---

Foundation: National Basic Research Program of China (973 Program), No. 2012CB955804.

Z. Wang et al., *Integrated Assessment Models of Climate Change Economics*, DOI 10.1007/978-981-10-3945-4_8

Programme (UNEP) have officially launched IPCC, the Intergovernmental Panel on Climate Change, which announced the Global Warming as a reality and calls for actions against it from all global communities. However, when it goes for concrete actions, the ethical debating surfaced and led to discussion on Ethics of the Climate Change (Gardiner 2004). In the recent years, the debating between some two dramatic figures, Nordhaus (2007) and Stern (2008), has escalated the international concerns on these issues. The Ethics of the Climate Changes containing the social and environmental ethics, with profound influences on the intricate personal and society well-being, deserves some systematic researches.

The crux of the Climate Ethics lies in the difference between science and humanity—science is neutral and objective in values while human beings are not. Under absolute neutral judgments, the Global Warming is nothing scaring but purely a technical issue. The reality is that the Climate Change has run deep into the intricacies of many social and political issues. Only when we realize why we have to do and make a global agreement according to their own value principle, can we solve the issue caused by climate change through the means of science and technology, and on this basis to discuss the responsibility for climate protection, the target of climate control, the reduction quotas of greenhouse gas emission, the national economic cost and action attitude moreover the procedural justice and other decision-making issues. This situation gives a prefect testimony to Voltaire's saying: *Ethics comes first in human needs*. All the countries on this globe have to answer this question. This article attempts to address on the Climate Ethics with the basic concepts, in association with the culture and geography of China.

8.2 The Significance of the Climate Ethics

Climate ethics issues can be traced back to the 1980s for and the 10th Conference of the Parties of the United Nations Framework Convention on Climate Change in December 2004 is a key point. At this meeting, due to the debate on the responsibility addressing on climate change, Rock Ethics Institute of Pennsylvania State University in the United States initiated a cooperative research program focusing on the ethical dimension of climate change, which released the "*Dimensions of Climate Change [EB/OL]*" in 2007. At the same time, Northcott, an internationally renowned ethicist from the University of Edinburgh, UK, published "*Climate ethics appeals to the world*" (Northcott 2007) appeal for international attention. These studies of climate ethics are always more concerned about the general environmental ethics. In fact, from the perspective of general ethics, human raise the issues of environmental ethics: What is the position of human-beings in the nature? What is the basic value of human-beings? What is our biological and social character respectively and what about its importance? What is the reasonable and valuable way of life for us to live in the nature?

At the Copenhagen conference of 2009, the response to climate change was revealed not only as the issues of environmental ethics, but also the moral issues of relationship between countries and nations at the height of "international ethics". After that conference, some British accused China of kidnapping the meeting while some Chinese accused developed nations of kidnapping the world. But now, the whole world has to pay for what they have done. These are forcing us to considering climate change as an ethical issue which challenges the human civilization but not only an environmental ethics. Many scholars from developing countries, such as Hao (2010), Lang (2010[1]), argue that the climate change has already evolved from a scientific issue into the social and political issue, becoming a new means for the developed countries to kidnap the world and maintain its existing global domination. Scholars from developed countries, such as Posner and Sunstein (2008), they want to put more emphasis on the responsibility of emissions reductions in developing countries, and defense for the negative treatment of greenhouse gas emission reduction of the United States under the name of justice.

In numerous studies at present, the core of climate ethics has been explained as that people in the same boat should help each other, or "In the Same Boat" Principle for short, namely human collaboration is needed to address climate issues. There is a quite popular fable reflects this viewpoint. Many tourists were in a sinking boat, only abandoning luggage could keep the ship to stay afloat. Therefore, we should not argue about whose luggage should be abandoned. This fable reflects the core viewpoint of the climate ethics: people in the same boat should just help each other, which sounds reasonable according to the traditional environmental ethics. As a result, the "climate chauvinism" is accused for stressing too much on individual country's interests. However, the developing countries have found a challenge to this theory, through their own development. The share of responsibilities among the countries is not the same as some western scholars stated. It goes beyond the ethics of environment into the ethics of social and economic development. The former story about the big sinking ship should be revised as: since there are differences in the wealth of the tourists, should the richer with much more luggage which lead to the broke down of the boat, take more responsibility. Further more, how to allocate luggage abandoning. If everyone has to abandon a piece of luggage, it means that the rich just have to abandon one pieced of their luggage while the poor has to abandon the only piece of luggage they have. For the poor, that means a complete loss of all their property, which would lead them to be frozen or starved. Is it fair on the ethics to ask everyone to abandon one piece of luggage equally? This raises the need for further research on climate ethics. Some people think that developed countries can gain more GDP with the same carbon emission due to their advanced technology. This idea is actually a climate colonial ideology, which means the developed countries are entitled to keep their advantages over the developing countries in sharing of the world resources. They completely forget, with the same

[1]Lang Xianping, Lang's Review on Finance and Economics: scandalous fraud of climate change, http://v.ku6.com/show/_c4ouaTrnnz4xnFSclnvZg.html?nv=1.

carbon emission, the increase of net benefits for developing countries may be less than that in the developed countries. But because of the initial welfare in developing countries is much lower than developed countries, the welfare improvement in developing countries is much prominent.

8.2.1 A New Perspective of Climate Ethics

Under this circumstance, what's the standing point of climate ethics? What is the norms, beliefs, attitudes and guidelines of human being's behavior which affect the change of climate? Some scholars have proposed standards for welfare evaluation, but humans' view on welfare carries on in stages, how to evaluate the long-term welfare, and how to determine the long-term is the key issues we have to settle. Since we believe that the climate change is an environmental issue, we need to go back to the environmental ethics. As an evolutionary selection of human existence, Chinese culture, a culture with long history, has a positive meaning to the establishment of climate ethics. There is a complete development of environmental ethics since ancient China (Wang and Wang 1998), and the archaic Chinese environmental ethics can be considered as the standing point of the climate ethics.

Chinese philosophy contends that human ethics requires the coexistence, which means harmony of human relations, named "He" in Chinese. The Chinese traditional ethical system has complete elaboration on this issue. As a kind of cultural choice when man evolved in the environment for survival, Chinese civilization with thousands of years of heritage is worth learning. Confucius said: "In practicing the rules of propriety, a natural ease is to be prized" (The Analects • Xue Er). This means that the core content the in implementing ethical system is "He" (means harmony). Chuang Tzu said "now in a high position and now in a low, he is in harmony with all his surroundings (Chuang Tzu, The Tree on the Mountain). Harmony has become a standard, because "Everything has been able to all-win and in health" (XunZi, The Tianlun). Finally, Dong Zhongshu concluded: "Harmony is greater than morals" (Dong Zhongshu, "ChunQiuFanLu"). This is a summary of the reality. In fact, "Harmony is greater than morals" is the evolutionary selection of human over the past millions of years. It is lucky for humans to choose harmony as principle to fight against natural risks, compose social cooperation to withstand disasters. Xunzi summarized the choice of human evolution, he said: "Why beasts of burden can be used by humans? It's because human can be combined into a social group, but they can not" ("XunZi, Wang Zhi"). Although the reason why beasts of burden become a tool of humans is not merely due to their lack of ability to cooperate, cooperation is indeed the selection of human's evolution.

Under the ethical principle of harmony, Chinese philosophy emphasizes the purpose of cooperation is to live together instead of profit competition. While the climate colonialism and climate chauvinism are in the opposite direction. Laozi deemed: "With all the sharpness of the Way of Heaven, it injures not; with all the doing in the way of the sage he does not strive." (LaoZi). The target principle of climate ethics is the cooperation for the co-existence, but not the maximization of welfare on temporary. And Xunzi said "If the principle to organize the social groups is appropriate, everything can get suitable arrangements, all animals can have its deserved growth, and all creatures could be able to gain its life." (XunZi, WangZhi). Everyone has the opportunity to obtain its deserved growth and lifespan. On the issue of how to achieve, Xunzi summarize the human evolutionary selection. He said "why beasts of burden become slaves for humans? Because humans can form social groups, but they can not. Why human can be combined to form a social group? That is because of the rank and grade. How to implement the rank and grade? It's because of the existence of morals. So after determining the rank and grade according to the morals, people will be able to gain the harmony and coordination in their life, which attribute to the solidarity. Power expands by solidarity, resulting prosperous, which can defeat external objects. Therefore, humans may live in their house. Therefore, the reason why people can order the four seasons to manage everything, and benefit the whole world is merely the rank and morals. (XunZi, Wang Zhi). Xunzi stressed the foundation to understand the ethics culture is the "rank" based on the "morals", here the "morals" is the standard of ethics, while the "rank" is the action principle of ethics. Xunzi Warning: "Humans can not live without social groups, but there will be contention if the humans combined into a social group without ranking and grading, which may generate unrest. Once the unrest happened, it will lead to the alienation of members and weaken the strengths, resulting the failing to defeat external objects." ("XunZi, Wang Zhi"). On ethics, the "morals" of Chinese philosophy means to understand each other. Confucius emphasized" The man of perfect virtue, wishing to be established himself, seeks also to establish others; wishing to be enlarged himself, he seeks also to enlarge others. "(The Analects, Yong ye). Voltaire highly praised this ethics thought of Confucius.[2]

It is worthwhile to note that the so-called "ranking the group" stressed by Xunzi focus on the allocation of the group's responsibility to individuals. And he also argues "All creatures could be able to gain its life." And "ranking the group" requires group interests should be implemented to individuals. Measurement of happiness and level of responsibility should be reflected in the collective as well as the concept of happiness and responsibility, which is a basic understanding of Chinese philosophy. As the common saying goes in China, "Every man alive has a duty to his country." It means that each individual should assume the responsibility for the collectivity, and this share is implemented into individual in the terms of public welfare. Confucius said, "I have heard that rulers of states and chiefs of families are not troubled lest their people should be few, but are troubled lest they should not keep their several places." ("The Analects, Ji Shi"). Xunzi said: "How to implement the rank and grade? It's because of the existence of morals. So after determining the rank and grade according to the morals, people will be able to gain the harmony and coordination in their life, which attribute to the solidarity. Power

[2]http://www.360doc.com/content/11/1124/03/7434782_166914190.shtml.

expand by solidarity, resulting prosperous, which can defeat external objects." (Xunzi, Wang Zhi) Morals mean justice and fairness. Implementing the welfare and the responsibility for emissions reductions by individual units on the issue of climate will be able to achieve the target through cooperation. Unfortunately, among large numbers of IAM (integrated assessment model), regarding the global welfare maximum as a criterion of emission reduction does not reflect the principle of "ranking based on morals", which directly leads to the idea of climate colonialism.

According to the sustainable human ethics of China which has been chosen by history in the long run, human need to identify the interests of the collective under the climate change. The measurement and point of happiness between interests collective are based on the individuals, namely "ranking the group", which requires to evaluate the collective interests on the basis of explicitness of individual interests. This is a new perspective of climate ethics. A comparison of emissions of OECD countries and non-OECD countries has been shown up in the IPCC report. It is quoted by the IPCC that emissions of non-OECD countries is much higher than that of OECD's, which seems that the non-OECD countries should take more responsibilities. In fact, the population of the non-OECD countries is far larger than those of OECD countries,and this classification for understanding the responsibility is the result of "inadvisable ranking", which can only bring about the contention and unrest. "Once the unrest happened, it will lead to the alienation of members and weaken the strengths, resulting the failing to defeat external objects." Human can not accomplish the goal of protecting the climate. Our simulation shows that taking no emission reduction is the Nash equilibrium among national gaming facing with climate change. No emission reduction of each country is a Nash equilibrium point when they feel the game.

It needs to be pointed out that the thought of "harmony" and "ranking the group" emphasize the ethics selection, which means everyone has the opportunity to develop its strengths and access to their own fate, should be traced back to the significance of welfare. As a result, welfare must go back to the individual. The importance of individual is that they are the true subject of emission reduction. It is difficult to imagine there would be the same choice of climate ethics between the children who go to school on barefoot and carry firewood, living in the edge of Nu River in Yunnan, China, and kids from Singapore wearing Nike shoes, sitting on a school bus, and holding the iPad. Because their opportunity and requirement to develop their strengths and access to their own fate are not the same. Where is the basis for ethics of climate protection for children without survival? If the emission reduction of China has been increased, how can the children from Nu River fulfill the emission reduction, whose per capita consumption of carbon is low. Why they should take on more losses than Singapore children? And their potential to achieve the emission reduction is very low, benefits of their emission reduction are not obvious. There is no doubt that the main responsibility for climate ethics should be traced back to the individuals rather than remaining at the national level.

If ignoring ethical assumptions, but only from the point of view of economics, there is a decreasing marginal value of money for each individual. This means that from a societal point of view, giving (or taking) an extra dollar to or from a poor person is equivalent to giving (or taking) and extra dollar to or from a rich person, but this is far from the truth. On the issue of emissions reducing, how much the poor and the rich should take the corresponding cost can cause unfair from the ethical perspective if only use the economics, because their opportunity and requirement to develop their strengths and access to their own fate are not the same and cannot get the uniform degree of satisfaction, even betray "Everything all-win and a living". How much of the cost of mitigation should be borne by the poor in comparison to what is borne by the rich? It is partly an issue of distributive justice. From the ethical perspective, the developed countries have the compensation liability to pay for the mitigation that takes place in poor countries.

Because of these new perspectives for climate ethics, we need to further explore the basic issue of climate ethics.

8.3 Basic Issues of Climate Ethics

8.3.1 The Equity Principle of Climate Ethics

The primary concern of climate ethics is whether our response behavior to climate change is fair. As we emphasized that the measurement and point of happiness among these interest collectives are established on the individual, the principle of climate ethics is the target of coping and ensuring the happiness of each individual. But among the controversy of the responsibilities sharing, we only consider the national interests rather than individual happiness and this problem had been completely politicized throughout the years. Actually, stressing equity on national level will harm the interests of citizens from the big countries, which is not fair. This scheme has led a pattern that countries pursuit their emission quotas and development opportunities, forming the basis of climate chauvinism.

Climate change mitigation demands large-scale action among various countries over the world, but all along, as to the discussion about emission reduction, economic analysis tells us that, for the sake of cost-effectiveness, the greatest reductions should be made where they can be made most cheaply; moreover, it also stresses that in the ideal case, emissions should be reduced in each place to just the extent that makes the marginal cost of further reductions the same everywhere. Obviously, countries and regions of the lowest reduction costs are developing and less developed areas. As mentioned above, when faced with emission reduction, both developed and developing countries should not shirk their responsibility, but the key point is totally different national conditions and the disparity of the economic situation makes the developed countries should shoulder more responsibility in emission reduction. Most of the anthropogenic Greenhouse Gas that is now in the atmosphere has been emitted by rich countries and much of the harm that is being done by these gases is suffered by people in poor countries. Developed countries have made significant economic development by the accumulation of historical emissions, however, the basic living conditions of the poor people in the poverty areas cannot be guaranteed. On this occasion, with the principle of "the greatest reduction should be made where they can be made most cheaply" certainly do not accord with Pareto criterion, and is also contrary to the "*Operation by Ranking and Grouping*" belief of this chapter.

Assessment on welfare is always adopted as the criterion for equity in the traditional climate ethics among which Ramsay function is cited as the basis for evaluation, for example, in the work of Nordhuas (2007), Stern (2008). In these efforts, maximized global welfare is taken as the basis for allocation of emission mitigation. In the study of WangZheng, etc. (Wang et al. 2012), he opposes the emission reduction allocation of maximizing global welfare which he believe will result in a more emission qouta for developed countries and lead to seriously unfair. In fact, the maximized global welfare can not bring about the convergence of global individual happiness and development opportunities in the future. Developed countries always have advanced technology, which can achieve more incremental welfare with the same amount of carbon emissions. Realizing maximized global welfare means that developing countries should give up the opportunities of carbon emissions to developed countries. It is typical climate chauvinism emphasizing that developing countries should sell the quota of carbon emissions to developed countries to increase global welfare.

In short, humans face dilemma on the issue of climate ethics. Humans can only identify the collective interests under the condition of climate change, and establish the happiness measurement and happiness point of the collective interests on basis of individual to avoid the climate chauvinism. On the other hand, humans need to oppose the principle of the maximizing global welfare cause this means the climate colonialism which contradicts to achieve the justice under the principle of "*harmony priority*" and "*Operation by Ranking and Grouping*".

To implement the justice, we need a measurable index. It seems that the Human Development Index from the report of UNDP can reflect the per capita equity or fairness. However, Sen (1985) emphasizes a new concept of happiness from the view-point of the Development Economics that the ability to obtain development is the fundamental index to measure happiness so that mitigating climate change should also improving the development. It requires that we should not only take the difference of happiness we enjoyed into account which means giving everyone the possibility to pursuit happiness, and also giving the opportunity for development. It also implies that we need the ethics of "*The man of perfect virtue, wishing to be established himself, also seeks to establish others; wishing to be enlarged himself, he also seeks to enlarge others*" *(Lun Yu,* "*Yong Ye*"*).*

Although this idea of sharing the development opportunities is mainly surging in developing countries at present, it also applies to the developed countries. If the amount of carbon emissions that the climate could burden is seen as the basis, Wang and Wu have calculated that developed countries have almost used up the assigned carbon emissions quota based on the principle of egalitarian in accumulated carbon emission per capita taking 1990 as the base year. From the standpoint of happiness emphasized by Sen, it is unfair not to give developed countries carbon emissions opportunity, which is also seen as "*Not morals*" from the perspective of traditional Chinese ethics. International community needs to recognize the ethics of "*The man of perfect virtue, wishing to be established himself, seeks also to establish others; wishing to be enlarged himself, seeks also to enlarge others.*" As the action ethics, Xunzi's thought of "*Operation by Ranking and Grouping*" is based on "*Ranking by the morals*". "Morals" is a principle of fairness and equity. It is unfair to require developing countries to stop the industrialization efforts to obtain development opportunities, while it is also unfair for developing countries to settle old accounts in future, denying developed countries should have the right of emissions, with which the Western will not agree. The genius of the Chinese ethics lies in "Ranking by morals". In the metaphor of wreck, the simple way of justice is that everyone abandons the same thing. However, the same piece of luggage to the value of each person's welfare is not the same. Therefore, we must discuss the issue of fairness on a broader level. We believe that in the terms of "*Ranking by morals*", by the standard of "*Everything has been able to all-win and in health*", and with the characteristic of "*Operation by Ranking and Grouping*", it usually means that the effect of emission reduction should be Pareto improvement. And after a certain period of time, for example till 2100, the welfare of all countries will be improved, which is the critical and viable basis for the fair emission reduction.

On the issue of equity or fairness, some countries or international organizations tightly pegged to the total carbon emissions of China and India, and intentionally or unintentionally overlook the high per capita consumption of carbon from some small countries, which is fundamentally unfair. However, according to the principle of absolute fairness which is based on the historical emissions, it means that they can continue to make substantial greenhouse gas emissions which will exacerbate the change of global climate. Faced with the warming of global climate, humans can only on "Ranking the group" on the basis of "He", considering the fairness of comprehensive happiness based on individual and national happiness and responsibility.

8.3.2 Justice and Responsibility

We have drawn clearly that "*Ranking by morals*" and "He" is the appropriate ethical principles which have been chosen in the long-term evolutionary selection, and human individuals and collectives enjoy equal rights on the happiness and welfare. To implement this kind of right, another basic starting point of the climate ethics being involved is justice, which requires scientific foundation including response and examination of the historical emissions and raising an issue of historical responsibility. At present, the fact is that the climate has already changed, and the developing countries have been driven by developed countries on the climate issue. Human-beings are facing with a series of events and the share of adaptation cost caused by climate change. So just considering historical responsibility can not solve the climate problem. It is a complex issue to realize the responsibility of all countries and all individuals to reduce the current and future emissions of greenhouse gases on the ethical basis of "morals", namely "*Everything all-win and a living*" which needs to be further back in ethics.

8.3.2.1 Causal Responsibility and Moral Responsibility

Humans benefit from the greenhouse gas emission, in which developed countries gain the most. Lamarque et al. (2010) pointed out, the process of industrialization in developed countries occurred in hundred years ago also attributes to the greenhouse effect and the change of climate. Therefore, they should assume considerable responsibility. Dong et al. (2012) has calculated this asymmetry that 2/3 of the responsibility for global warming comes from developed countries but some popular mitigation schemes require developing countries to take 2/3 of the task of emissions reduction. However, as IPCC (2007) alleged, developing countries have relatively smaller historical responsibility but more potential CO_2 emissions in the future which should be a greater risk to future climate change. Nowadays, developing countries, with demand of economic development and relatively backward technology and knowledge, will inevitably bring about greenhouse gas emissions which will have impact on climate change in the future. Obviously the impact between the two parts is asymmetric. We should understand that the asymmetry here is the history which cannot turn back, greenhouse gas emission reduction is the real necessary which we can not avoid.

This illustrates that the causal responsibility of historical emission factors can not be ignored in determining the global responsibility. And due to the non-symmetry of the issue, from the perspective of the morals of "He", the developing countries have a responsibility for emissions reductions and need to be involved in the emission reduction action. The problem here is that the causal responsibility can not be thus end. In the terms of "*Ranking by morals*" and with the characteristic of "*Operation by Ranking and Grouping*", those needed to assume causal responsibility should take on the compensation obligation. On the previous Shipwreck story, after the poor abandoned their last rags except last luggage, the rich have the responsibility of giving their warm clothes to the poor, which is the completion of responsibility for the developed countries. In the Durban climate conference, it is pledged that developed countries will provide funds for climate protection. This is a good starting point, however, it is not enough. From the view-point of ethics, the developed countries should assume compensation responsibility at the first place, which not only means the investment in capital, but also reflect in terms of technology transfer, education improvement and social relief, instead of regarding the reducing of greenhouse gas emissions as the market opportunity for some country. In fact, the developed countries have become the welfare state by the early development. From the perspective of causal and moral responsibility, developing countries should enjoy tolerance in order to protect their individual development opportunities which mean the last chance for developing countries to realize the "*Ranking by group*" and "*Everything all-win and a living*", and to accomplish the shared responsibility of developed countries to help developing countries to complete responsibility for emissions reductions. In this regard, ethics improvements is not enough, we also need some legal protection, which gives ethical basis for developing countries to get what they deserve according to the terms of individual average.

8.3.2.2 Justice of Quotas

Just the causal responsibility and compensation responsibility is not enough, climate protection requires all humanity to share responsibility. "Ranking the group" first proposes a fair quota of greenhouse gas emissions, namely how we allocate the quota to countries and regions fairly. There are also a lot of controversies on the adoption of allocation method. First of all, whether we regard the individual fairness as a starting point, however, will not work in a general sense. If we trace back to the historical responsibility of developed countries, such as the United States, the United Kingdom, and even Japan, they have exhausted their deserved quota for a long time. It is clear that the developed countries still need the right to pursuit their economic growth, which is acknowledged by the ethics of "He". At present, some countries insist on the fairness of per capita emission, and they think the allocation should be tilted to some extent, by providing the poorest of the population with the greatest emission right and giving minimal emissions to the most affluent population. The drawback of this approach is that the emissions resources would be apparently wasted, because the technology in the poorest countries are often out-dated, leading the stagnation of world economy, which is extremely unreasonable from the judgment principle of value. Therefore, the most acceptable is to establish universal happiness standard, which is enjoyed by all human beings. Under the preconditions of investment is certain, we should give full consideration to the maximum utility or "welfare". The quota allocation emphasizes the justice instead of fairness.

In the actual operation of the justice standard, we must distinguish between two choices: First, proportional allocation for emission rights, to equally distribute the emission rights in accordance with the proportion of the population; the other is the prioritized allocation for emission rights. Both of the two method are confronted with controversy of varying degrees.

According to the first principle, the global emission reduction would be faced with the challenge: the costs and benefits of greenhouse gas emissions are quite different in various regions of the world and the various ethnic groups. And the happiness or the so called welfare brought about by the emissions of greenhouse gas would also be very far apart from each other. Therefore, this allocation method is also difficult to reach an agreement. Let's take China as an example. Xiao and Zheng (2013) found by calculation that, taking the future requirements of the carbon emissions into consideration, the province with the largest surplus is Sichuan Province, followed by Henan Province. While the provinces with largest gap is Shanxi Province, followed by Shandong, Hebei, Liaoning, Zhejiang,

Shanghai, Inner Mongolia etc. It is easy to find that the economic core areas of China will be suppressed and the energy basis can not get guarantee for its development under such conditions, probably leading the turbulence for the economy of China. However, the second allocation principle is clearly strengthening the unfairness. We take China as the example. The calculations of Xiao and Zheng (2013) show that Shanghai can obtain the largest amount of emissions quota under the allocation principle in accordance with the GDP, followed by Beijing. It may appear the situation that the rich get richer, the poor get poorer, which is obviously unfair, and the result is unfair.

A case of the second principle of distribution is the "Color preservation" (Yang and Sirianni 2010). The basic idea for this principle is to allocate carbon qouta based on the current proporation of carbon emission. The advantages of this idea is to help to maintain the existing structure of national development, without bringing new shocks to the world economy, which is feasible in the development economics. But the principle of Yang and Sirianni is the result of a global welfare Optimal Gaming, its drawback is that it does not reflect the rapid development of developing countries and the viewpoint of the development economics that right to develop is a kind of happiness, which needs improvement.

Usually we considered that the distribution of the ratio of greenhouse gas emissions should be tilted to the less developed countries and regions. Suppose we have an effective and fair global carbon trading market, the differences among countries and regions will be correspondingly reduced (by carbon trading). This distribution depends on how to fairly and effectively determine the amount of the initial allocation, which also needs a scientific computing. Not very optimistic, there is little quota can be allocated to the backward countries and regions. It is likely that the gap between them and the developed countries will continue to exist, which will not be narrowed. The justice of allocation can not be guaranteed. Thus we must take a clear stand on the allocation of carbon quota to tilt to the underdeveloped countries sometimes.

8.3.2.3 Intergenerational Justice and the Rights of Future Generations

Intra-generational justice is concerned in climate ethics as it is concerned in general sustainable development issues. According to the theory of the sustainable development, 遇 we think intergenerational justice includes contemporary responsibility to future generations as well as the responsibility to Contemporary generation (Rawls 1971). Relating to the effect of greenhouse gas emissions on the quality of future generations' life, some scholars, such as Caney (2006), Meyer and Roser (2009), Wolf (2009) introduced the following ideas on the level of emissions:

(5) Future people's basic rights include rights to survival, health and substance;
(6) These basic rights–often understood as a subset of the general and universal moral rights whose protection is owed to all by all, i.e. as a subset of human rights–are very likely to be violated when temperatures rise above a certain level;

(7) Currently living people can slow the rise in temperature by limiting their emissions, and they can do so at reasonable costs to themselves; thus,
(8) A reduction in emissions is required for currently living people to fulfill their minimal duties of justice to future generations.

Clearly, the intergenerational justice of climate stresses that past generation, contemporary generation and the future generation have the common right of survival, life and development. But when they enjoy the climate resources, they should also take responsibility for maintaining the quality of the climate resources at the same time. If contemporary generation consume and reduce the quality of climate resources, they ought to make reasonable compensation to future generations. In fact, intergenerational justice issues related to climate change are often used as major weapons to remove obstacles in climate negotiations (Wang and Xu 2011).

However, it seems that this easy to be recognized issue has caused a great deal of dispute originated from the assessment of the long-term effects of climate change. In the analysis of the economics of climate change, the loss of economic welfare caused by climate change is defined as Keynes–Ramsey utility function which means: the utility is a function of the per capita consumption $C_{i,t}/L_{i,t}$. In particular, consumers usually consume with preference, namely there are "short-sighted". Therefore utility changes over time, there is a time preference, which is defined as the rate of depreciation. Assuming that the relative risk coefficient of a unit of consumed currency reflected by every consumer representative is τ, then we get the following:

$$U(n) = \sum_{i}\sum_{t=1}^{n} \exp(-\rho_i t) L_{i,t} \frac{(C_{i,t}/L_{i,t})^{1-\tau}}{1-\tau} \tag{8.1}$$

Here, $C_{i,t}$ means the consumption at the stage of t, consumer groups $\{C_{i,0}, C_{i,1}, \ldots\}$ are subject to production resource constraints, i is the regional or national ID, n represents the number of the age to be considered. According to the Economic theory (Barro and Sala-i-Martin 2004), two parameters of ρ_i and τ reflect the structure of consumer social welfare, which eventually manifested in the depreciation rate of the capital. In recent years, climate economists dispute a lot around the value of depreciation rate, sparking a fierce ethical debate. President of the American Economic Association, Nordhaus (2007) from Yale University, deemed that the value should be 0.015, with the equivalent depreciation rate of 5%. But the Lord of United Kingdom, Stern (2008) from the London School of Economics considered it should be 0.001, with the equivalent depreciation rate of 1.4%. Thus, Stern calculated that emission reduction should begin immediately, and required developing countries to extend the emission reduction, otherwise the global economic will suffer great losses although at the same time Nordhaus gave another conclusion that developing country can start emission reduction later.

In fact, the economic growth of United Kingdom now is very slow after early long term development which leads to the very low discount rate. On the same time,

U.S. is still in a thriving growth period with faster discount rate. As for China, the discount rate has been estimated at 9% if still need to keep the rapid economic growth. From this way, Interests of future generation is reflected to the contemporary generation but there is still a big difference ρ_i among different countries and contemporary people in developing countries suffer extreme pressure. Therefore, we suggest recognizing the interests of future generation on ethics first, and ensure the current generation can survive especially in poor countries. No matter how to reduce the emission, no future generation without contemporary generation's living. Stern's point of view has been widely accepted by the majority of people in European politics, but the ethics basis of Chinese "morals", namely "*Everything all-win and a living*" tend to make people adopt the viewpoint of Nordhaus. We reckon that Nordhaus' standpoint is more in line with the reality of the world. The old adage says: "*The children have their own blessing*", because children and grandchildren will create wealth on their own. With economic growth, the welfare of the future may be created.

8.4 Justice of the Climate Negotiations

To achieve a global climate governance framework, rounds of negotiations and conferences have been held under the sponsor of the United Nations. In pursuit of the stipulations on emission quotas, the ethical justice is on demand and has entered the focus of the negotiations, where procedural justice is required. The fundamental proposition of procedural justice is that the decisions shall be made through a fair, public consultation and independent from the executive parties. It is too early to determine that the procedural justice has been secured in the global climate negotiations, as it has been hard for the developing countries to have sufficient numbers of social entities to representative the opinions of the majorities due to their relatively less developed economic status. Furthermore, the rapid growth in these developing countries has constituted challenges to the interests of the existing commercial interests groups, who are equipped with social powers to lobby the climate negotiations in favor of their commercial interests. Thus, some rules need to be established for the climate negotiations. The fairness in deed may not be secured by the procedural justice alone, as the developed countries inherit their predominance in the world economic affairs in modern history, which offers them advantages over the developing countries in the international organizations to manipulate the global climate stipulations.

To proceed on the Justice in Climate Negotiation, it is proper to look into the characteristics of it. The nature of the decision making participants is the first factor which will affect the justice of climate negotiations. In the climate negotiations, the decision makers are the sovereignty countries, with equal rights to each of them. It is this equal distribution of rights which leads to an actual unequal power sharing in the consideration that the ethical value judgment is actually based on the interests of each individual. The BRIC countries are often heavily blamed at the international climate discussion, regardless of their weights in world population. Many talks ended up without any conclusions unpleasantly, as some unions of countries small in population exploit the rights of the heavily populated countries. Such situations can only be improved by the awareness of "the ethical value judgment is based on the interests of each individual. Therefore, it is reasonable to set the principle of "operation by ranking and grouping" on the individuals at the core of negotiation rules for international climate negotiations. This is to ensure that each country will be treated equally. Many international incidents have proven that the residents of a country will be driven into a worse condition of equity and justice when the country itself is deprived of the equity and right it deserves. The equality among the countries is rooted from the equality that can be traced down to all the individuals.

Under the theory of that the ethical value judgment should be based on the interests of each individual, the most reasonable negotiation principle is to take all influenced individuals into the process of the policy making (Whelan 1983). However, this may turn out to be too extensive to be practical. Some scholars, like Caney (2005), Dryzek (2010) suggest that the NGOs and some social activists may representative the residents better in the policy making process. But these "book worms" get indulged in Utopian ideas from time to time, leading the negotiation into empty endings. It is even worse that the presentation of individual value judgments is more from the developed countries as the NGOs are poorly developed in the developing countries due to various economic and political reasons. On the other side, the international community may be put into the turmoil of populism when the governments in the developing countries advocate the NGOs on large scales. The countries in fever of populism often obstacle the international climate talks from any constructive results. Therefore, the negotiation rules need to be established for the international community. The UN Security Council, consisting of 5 permanent members in association with non-permanent members, has set a good example of applying the practice of "operation by ranking and grouping", for stipulations on important world issues. This may also work well in international climate negotiations.

The Climate Change affects everyone's life and the counter-measures require the collaboration of everyone. Any breach of procedural justice, which imposes the depression on the majority of the residents on this planet, will certainly jeopardize the people's participation. On the contrary, a good performance of procedural justice may educate the people about the severity of the Climate Change and necessity of climate governance (protection?). Their expressions and participation may also facilitate the framework on the global climate governance.

8.5 Conclusion

Active responding to the Climate Change is a common goal and mission of all human beings. However, any concrete measures require the economic and ethical considerations. As discussed previously in this article, Climate Change is not only a subject of science but also a subject in ethics research. It is not only a topic in methodology but also a topic in cognition study. The better approach of scientific researches on the Climate Change should integrate the studies on economic realities as well as ethics. Based on its primary discussion on climate ethics, this article introduced a climate ethical principle derived from Chinese traditional ethics:

On the value judgment in Climate Ethics, the responsibilities and rights should be shared according to the ranking and grouping. The emission quotas should be fairly distributed with reference to the development needs of each country. The reduction of emissions shall not hurdle the Pareto Optimality of each participating country. The foundation of the climate ethics lies in the awareness that the collective interests of a group is based on the individual interests in the group. People need to distinguish the parties of interests under the influence of the Climate Change, along with the wellbeing and measures of happiness of each party. The judgments on values should be made on the basis of the individual benefits, with regard to the Climate Change.

Furthermore, the article discussed about the Climate Ethical Criteria—Equity/Justice and Values, and analyzed the ethical definitions of justice and values in Climate Ethics. Thus, it proposed the Justice of Climate Ethics should be realized in distinguishing the causal and moral factors in the sharing of responsibilities, the stipulation of fair quotas and the integration of the interests among the existing and upcoming generations. The article explicated the justice in the climate negotiations in these captioned aspects and suggested that some climate proposals from the developed countries are inappropriate in ethics.

In short, we suggest an international ethical principle on climate issues: all the positive measures on climate governance or against the Climate Change should promote the economic and social development, as well as narrowing the gaps among the countries, without depressing the social individuals which are being poor deeper into poverty. The equity and justice for every individual should be on the focus of the international climate policy-making. The residents of a country will be driven into a worse condition of equity and justice when the country itself is deprived of the equity and right it deserves. Any such climate governance policy or measures against the Climate Change will certainly hurdle the willingness of these people to participate, and thus non-practical nor sustainable.

References

Barro R, Sala-i-Martin X (2004) Economic growth. MIT Press, Cambridge

Caney S (2005) Cosmopolitan justice, responsibility, and global climate change. Leiden J Int Law 18:747–775

Caney S (2006) Cosmopolitan justice, rights and global climate change. Can J Law 33 Jurisprud XIX:255–278

Wei T, Dong WJ et al (2012) Developed and developing world responsibilities for historical climate change and CO2 mitigation. In: Proceedings of the national academy of sciences of the United States of America. doi:10.1073/pnas.1203282109/-/DCSupplemental

Dryzek J (2010) Foundations and frontiers of deliberative governance. Oxford University Press, Oxford

IPCC (2007) Synthesis report. Cambridge University Press, Cambridge, 52 pp. http://www.ipcc.ch

Jamieson D (1992) Ethics, public policy, and global warming, science, technology. Hum Value 17(2):147

Lamarque JF, Bond TC, Eyring V et al (2010) Historical (1850–2000) gridded anthropogenic and biomass burning emissions of reactive gases and aerosols: methodology and application. Atmos Chem Phys 10:7017–7039

Meyer L, Roser D (2009) Enough for the future. In: Gosseries A, Meyer L (eds) Intergenerational justice. Oxford University Press, Oxford; New York, pp 219–248

Nordhaus WD (2007) A review of the stern review on the economics of climate change. J Econ Lit XLV:686–702

Northcott MS (2007) A moral climate: the ethics of global warming. Darton Longman and Todd, London

Posner EA, Sunstein CR (2008) Climate change justice. Georgetown Law J

Rawls J (1971) A theory of justice. Belknap Press, Cambridge

Rose A (1998) Burden-sharing and climate change policy beyond kyoto: implications for developing countries [J]. Environ Dev Econ 3(3):347–409

Sen A (1985) Commodities and capabilities. Oxford University Press, Oxford

Skrimshire S (ed) (2010) Future ethics: climate change and apocalyptic imagination. Continuum International Publishing Group, London

Gardiner SM (2004) Ethics and global climate change. Ethics 114

Stern N (2008) Key elements of global deal on climate change. The London School of Economics and Political Science (LSE)

Wang Z, Zhang S, Wu J (2012) A new RICEs model with the global emission reduction schemes. Chin Sci Bull 57(33):4373–4380

Whelan F (1983) Prologue: democratic theory and the boundary problem. In: Pennock JR, Chapman JW (eds) Nomos XXV: liberal democracy. New York University Press, New York, pp 13–47

Wolf C (2009) Intergenerational justice, human needs, and climate policy. In: Intergenerational justice. Oxford University Press, Oxford, pp 347–376

Yang Z, Sirianni P (2010) Balancing contemporary fairness and historical justice: a 'quasi-equitable' proposal for GHG mitigations. Energy Econ 32:1121–1130

Voltaire (1990) Essay on the manners of nations. In: Chinese translation. The Commercial Press, Beijing

Voltaire (1991) Dictionnaire philosophique. In: Chinese translation. The Commercial Press, Beijing, p 343

Xiao L, Zheng W (2013) Regional carbon emission control in China. In: Zheng Wang (ed) Carbon emission control strategies research in China. Science Press, Beijing

Hao Q (2010) Justice, rights and responsibility: ethical considerations on the issue of climate change. World Econ Polit 10:58–72

Wang S, Xu F (2011) Climate justice: how to be possible and its principles. Janghai Acad J [J] (3): 130–135

Wang Z, Wang L (1998) The principles of environmental ethics of sustainable development. In: Juemin X (ed) The study of human geography in China. Science Press, pp 216–227

Jing W, Zheng W (2009) Global emissions: scheme analysis and key problems. J Chin Acad Sci 24(5):475–485